# KiCad Eeschema Reference Manual

A catalogue record for this book is available from the Hong Kong Public Libraries.

Published by Samurai Media Limited.

Email: info@@samuraimedia.org

ISBN 978-988-13277-8-9

# Copyright

This document is Copyright © 2010–2013 by its contributors as listed below. You may distribute it and/or modify it under the terms of either the GNU General Public License (*http://www.gnu.org/licenses/gpl.html*), version 3 or later, or the Creative Commons Attribution License (*http://creativecommons.org/licenses/by/3.0/*), version 3.0 or later.

All trademarks within this guide belong to their legitimate owners.

## Contributors

Jean-Pierre Charras, Fabrizio Tappero.

## Feedback

Please direct any comments or suggestions about this document to the kicad mailing list: *https://launchpad.net/~kicad-developers*

## Acknowledgments

None

## Publication date and software version

Published on March 7, 2014.

## Note for Mac users

The kicad support for the Apple OS X operating system is experimental.

# Table of Contents

# 1 - Introduction to Eeschema

## Table of Contents

## 1.1 - Description

Eeschema is powerful schematic capture software distributed as part of KiCad and available under the following operating systems:

- Linux.
- Apple OS X (experimental).
- Windows XP, Windows 2000, Windows 7.

Regardless of the OS, all Eeschema files are 100% compatible from one OS to another.

Eeschema is an integrated software where all functions of drawing, control, layout, library management and access to the PCB design software are carried out within Eeschema itself.

Eeschema allows the use of hierarchical drawings using multi-sheets diagrams. Eeschema handles:

- Flat hierarchies.
- Simple hierarchies.
- Complex hierarchies.

Eeschema is intended to work with printed circuit software such as PcbNew, for which it can provide the Netlist file, which describes the electrical connections of the PCB.

Eeschema also integrates a component editor which allows the creation, editing, and visualization of components, as well as the handling of the symbol libraries (Import, export, addition and deletion of library components).

Eeschema integrates the following additional but essential functions needed for modern schematic capture software:

- Design rules check (DRC) for the automatic control of incorrect connections and inputs of components left unconnected.
- Generation of layout files in POSTSCRIPT or HPGL format.
- Generation of layout files printable via local printer.
- Bill of Material generation.
- Netlist generation for PCB layout or for simulation.

## 1.2 - Technical overview

Eeschema is limited only by the available memory. There is thus no real limitation to the component count, number of component pins, connections, sheets. Eeschema allows simple or multi-sheet diagrams.

In the case of multi sheets diagrams, the representation is hierarchical, and the access to each sheet is immediate.

Eeschema can use multi-sheet diagrams of this type:

- Simple hierarchies (each diagram is used only once).
- Complex hierarchies (some diagram is used more than once, multiple instances).
- Flat hierarchies (some diagrams not explicitly connected in a master diagram).

The maximum size of the drawings is always adjustable from A4 format to A0 or from A to E format.

# 2 - Generic Eeschema commands

## Table of Contents

## 2.1 - Access to Eeschema commands

You can reach the various commands by:

- Clicking on the menu bar (top of screen).

- Clicking on the icons on top of the screen (general commands).

- Clicking on the icons on the right side of the screen (particular commands or "tools").

- Clicking on the icons on the left side of the screen (display options).

- Clicking on the mouse buttons (important complementary commands). In particular a right click opens a contextual menu that depends on the element under the cursor (Zoom, grid and edition of the elements).

- Function keys of the keyboard (F1, F2, F3, F4, Insert and space keys).
  Specifically:
  The "Escape" key often allows the canceling of a command in progress.
  The "Insert" key allows the duplication of the last element created.

Here are the various possible accesses to the commands.

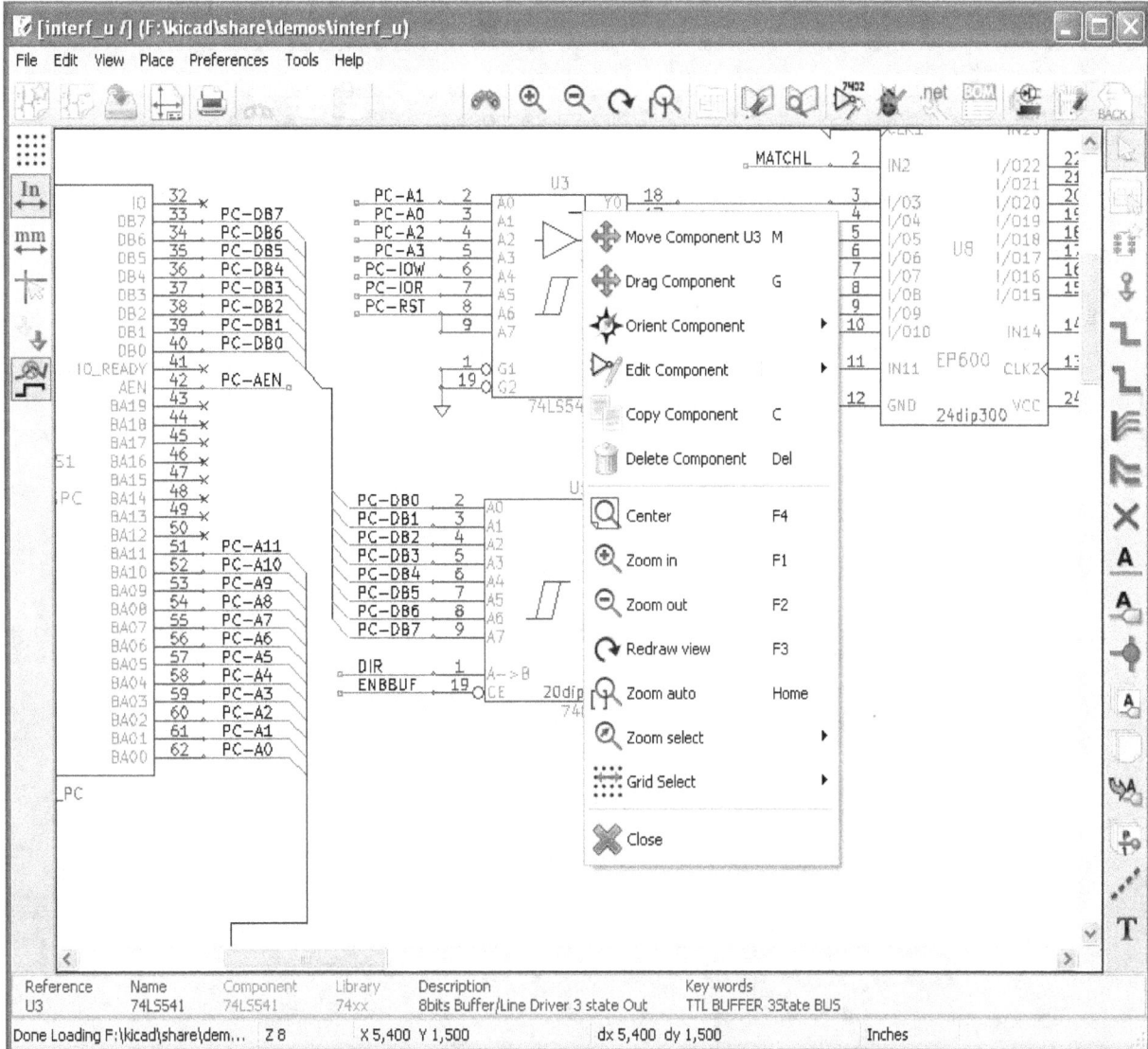

## 2.2 - Mouse Commands

### 2.2.1 - Basic commands

**Left button**

- Single click : displays the characteristics of the component or text under the cursor.
- Double click: edit (if the element is editable) the component or text.

**Right button**

- Opens a pop-up menu.

### 2.2.2 - Operations on blocks

You can move, drag, copy and delete selected areas in all Eeschema menus.

Areas are selected with the left mouse button. The command is completed with the release of the button.

By holding one of the keys "Shift", "Ctrl", or the 2 keys "Shift and Ctrl", during selection this results in the copying, dragging or deletion of the selected area.

Commands summary:

| left mouse button | Move selection. |
|---|---|
| Shift + left mouse button | Copy selection. |

| left mouse button | Move selection. |
|---|---|
| Ctrl + left mouse button | Drag selection. |
| Control + Shift + left mouse button | Delete selection. |

The command is executed at button release.

During selection you can:

- Click again to place back the elements.
- Click the right button to cancel.

If a move block command has started, an other command block can be deselected via the pop-up menu (mouse, right button):

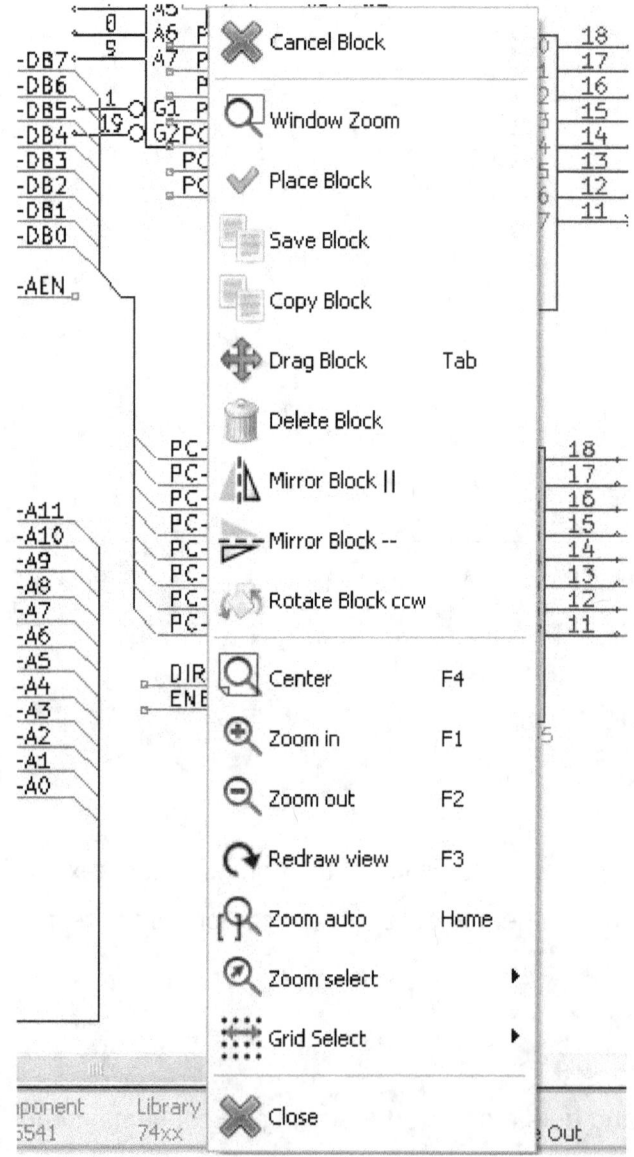

## 2.3 - Hot keys

The hot keys are not case sensitive.

- The "?" key displays the current hot keys list.
- The Preference menu manage the hot keys.

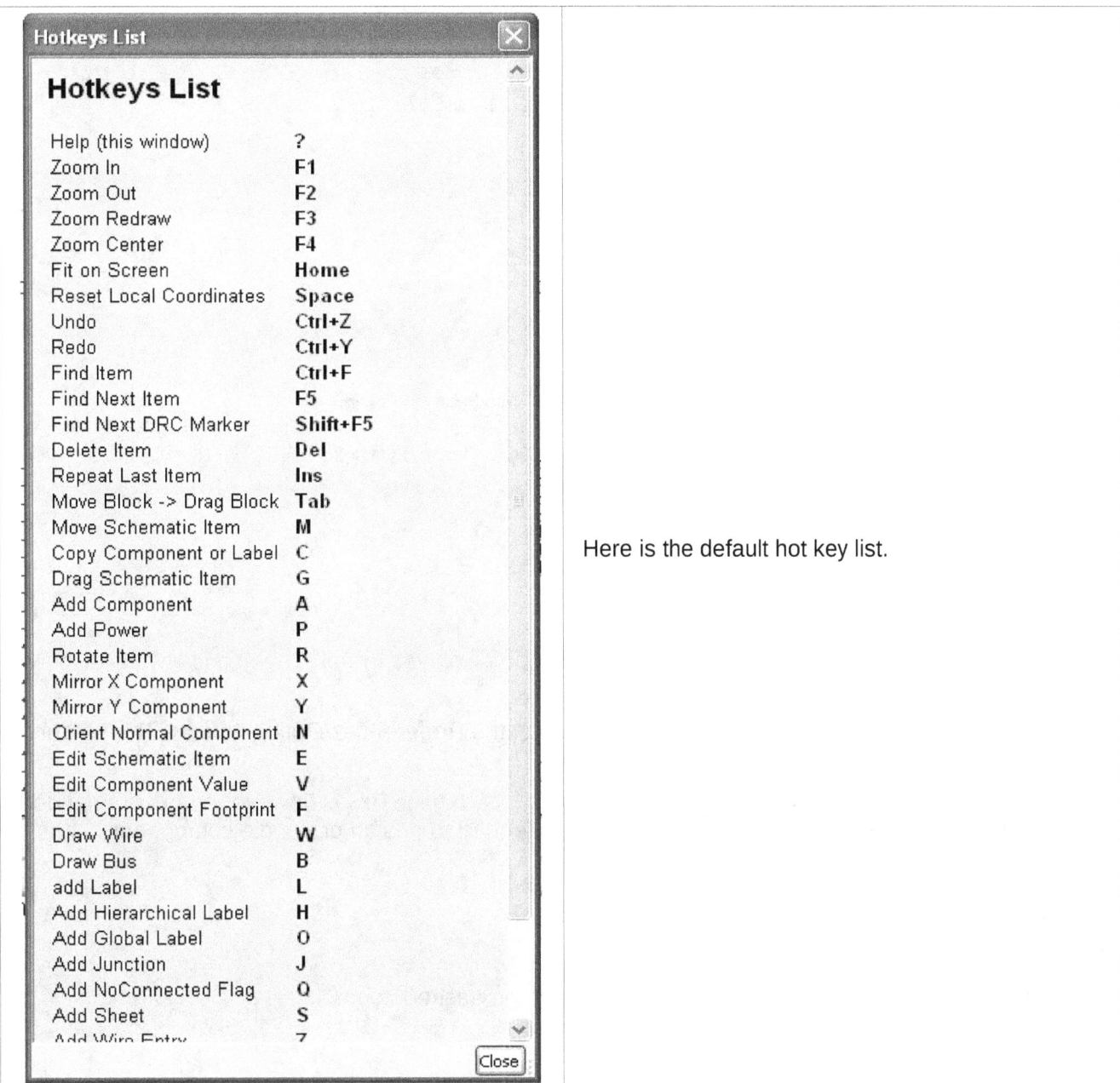

Here is the default hot key list.

All hot keys can be redefined by the user via the hotkey editor.

## 2.4 - Selecting grid size

In Eeschema, the cursor moves over a grid, which can be displayed or not. The grid is always displayed in the library management menus.

You can change the grid size via the pop-up menu or via the Preferences/Options menu. The default grid size is 50 mil (0.050 ") or 1,27 millimeters.

One can also work with the average (20 mil) or a finer grid (10 mil). This is however not recommended for usual work. The average or fine grid is especially intended to design or handle components with large numbers of pins like several hundreds pins.

## 2.5 - Zoom selection

To change the zoom level:

- Right click to open the Pop-up menu and select the desired zoom.
- Or use the function keys:
    - F1: Zoom in.
    - F2: Zoom out.
    - F3: Redraw.
    - F4: Center around the cursor
      Or simple click on the mouse middle button (without moving the mouse)
- Window Zoom: Mouse drag, with the middle button.
    - Mouse weel: Zoom in / Zoom out
    - SHIFT+Mouse weel: Up/down panning
    - CTRL+Mouse weel: Left/Right panning

## 2.6 - Displaying cursor coordinates

The display units are in inches or millimeters. However, Eeschema always works internally with 1/1000 of an inch.

The following information is displayed at the bottom right hand side of the window :

- The zoom factor.
- The absolute position of the cursor.

- The relative position of the cursor.
- The relative coordinates (x, y) can be reset with the space bar.
- The coordinates posted will then relate to this point.

## 2.7 - Top menu bar

The top menu bar allows the opening and saving of schematics, the program configuration, and it also contains the help menu.

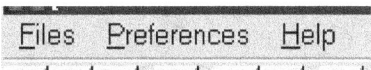

## 2.8 - Upper toolbar

This toolbar gives access to the main functions of EESchema.

| | |
|---|---|
| | Create a new schematic. |
| | Open a schematic. |
| | Save complete schematic (with the whole hierarchy). |
| | Select the sheet size and title block editing. |
| | Open print menu. |
| | Remove the selected elements during a move block. |
| | Copy selected elements in the clipboard during a move block. |
| | Copy last selected element or block in the current sheet. |
| | Undo: Cancel the last change (up to 10 levels). |
| | Redo (up to 10 levels). |
| | Call the menu of components localization and texts. |
| | Zoom in and out, around the center of screen. |
| | Redraw of the screen and optimal Zoom. |
| | Call the navigator window, to display the tree structure of the diagram hierarchy (if it contains sub sheets) and the immediate selection of any sheet of the hierarchy. |

| | |
|---|---|
| | Call component editor <u>Libedit</u> (Examination, modification, and editing of library components). |
| | Display libraries (Viewlib). |
| | Component annotation. |
| | ERC (Electrical Rules Check). ERC automatically checks for electrical connections. |
| | Creation of the netlist (Pcbnew , Spice format and other formats). |
| | Generate the BOM (Bill of materials) and/or hierarchical labels. |
| | Call CVPCB. |
| | Call PCBNEW. |
| | Import a stuff file from Cvpcb (fill the footprint field of components) |

## 2.9 - Right toolbar icons

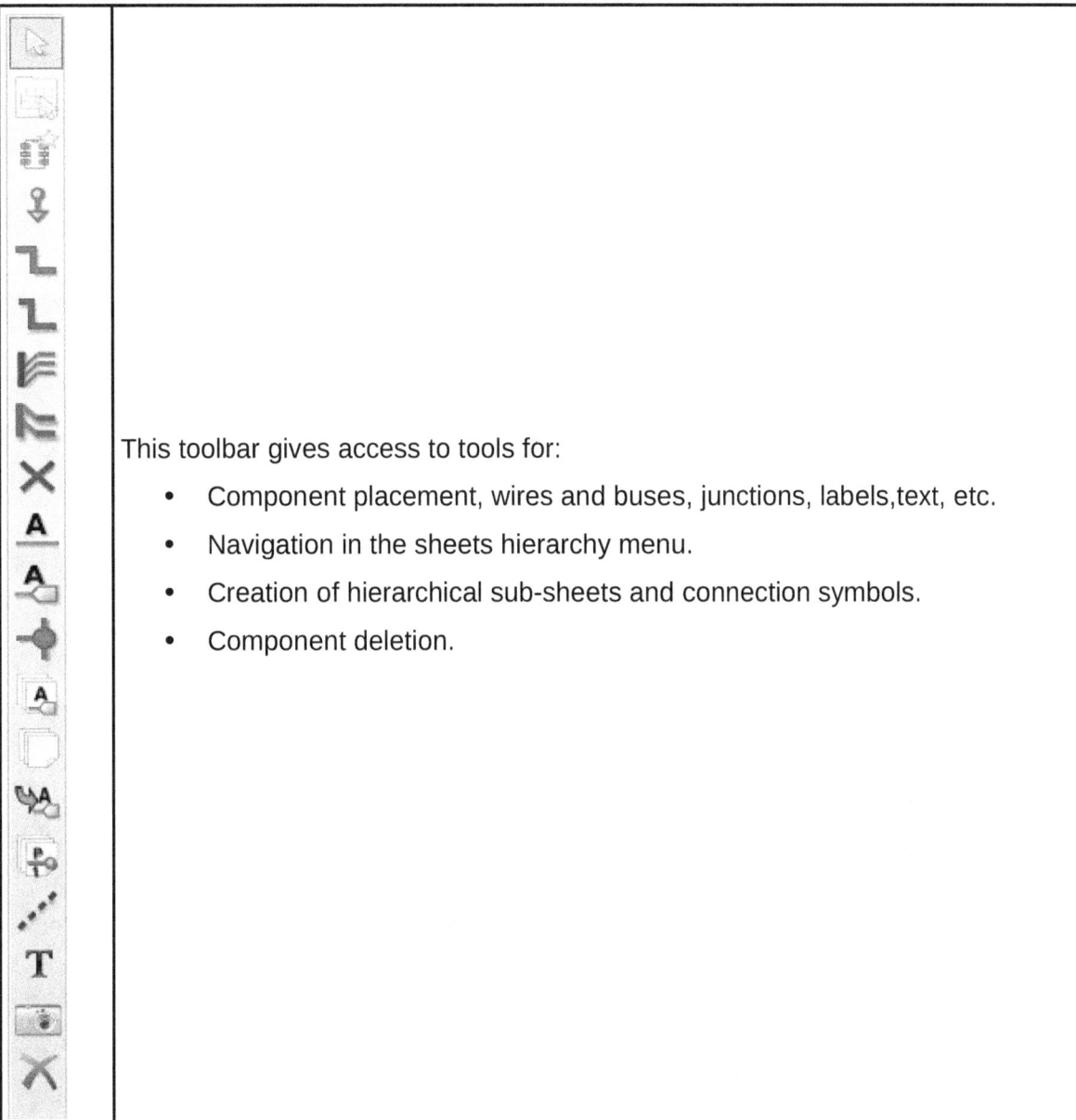

This toolbar gives access to tools for:

- Component placement, wires and buses, junctions, labels,text, etc.
- Navigation in the sheets hierarchy menu.
- Creation of hierarchical sub-sheets and connection symbols.
- Component deletion.

The detailed use of these tools is described in the chapter " Diagram Creation/Editing". An outline of their use is given below.

| | |
|---|---|
| | Stop the order or tool in progress. |
| | Navigation in the hierarchy: this tool makes it possible to open the subsheet of the displayed schematic (click in the symbol of this subsheet), or to go back up in the hierarchy (click in a free area of the subsheet) |
| | Call the component placement menu. |
| | "Powers" placement menu. |
| | Wire placement. |
| | Bus placement. |
| | Wire to bus connections. These elements have only a decorative role and do not allow connection; thus they should not be used for connections between wires. |
| | Bus to bus connections. They can only connect two buses between themselves. |
| | "No connection" symbols. These are to be placed on component pins which are not to be connected. This is useful in the ERC function to check if pins are intentionally left not connected or are missed. |
| | Local label placement. Two wires may be connected with identical labels **in the same sheet**. For connections between two different sheets,you have to use global symbols. |
| | Global label placement. All global labels are connected (even between different sheets). |
| | Junction placement. To connect two crossing wires, or a wire and a pin, when it can be ambiguous. (i.e. if an end of the wire or pin is not connected to one of the ends of the other wire). |
| | Hierarchical label placement. This makes it possible to place a connection between a sheet and the root sheet which contains this sheet symbol. |
| | Hierarchical subsheet symbol placement (resizable rectangle). You have to specify the file name to save the data of this "subsheet". |
| | Global label importation from subsheet, in order to create a connection on a subsheet symbol. Global labels are supposed to be already placed in this subsheet. For this hierarchy symbol, the created connection points are equivalent to a traditional component pin, and must be wired. |
| | Global label creation in subsheets to create connection points. This function is similar to the previous one which does not require already defined global symbols. |
| | Lines for framings… Only decorative, and does not perform a connection. |
| | Placement of comment text. Only decorative. |
| | Insert à bitmap image. |
| | Delete selected element. if several superimposed elements are selected, the priority is given to the smallest (in the decreasing priorities : junction, NoConnect, wire, bus, text, component). This also applies to hierarchical sheets. Note: the "Undelete" function of the general toolbar allows you to cancel last deletions. |

## 2.10 - Left toolbar icons

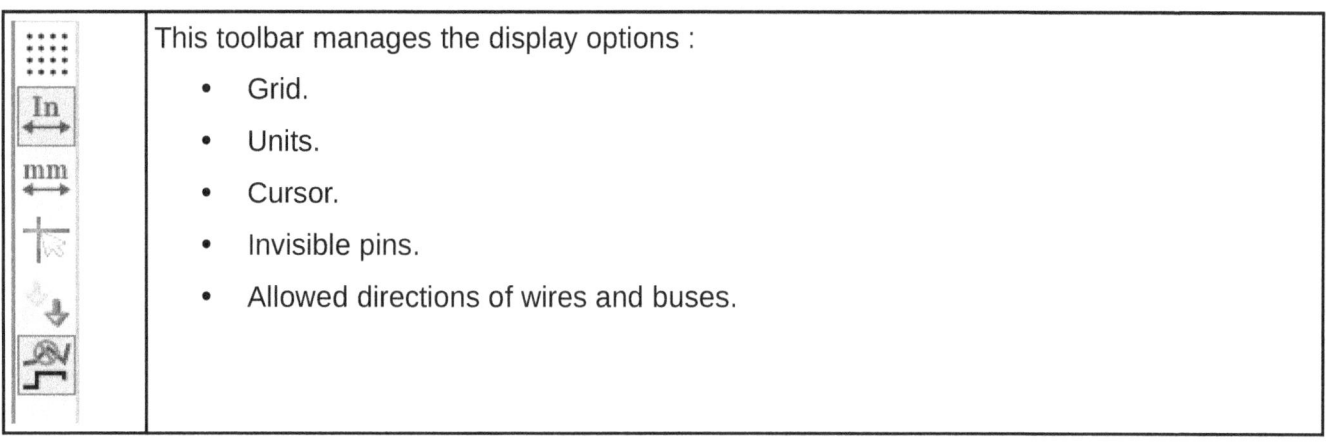

This toolbar manages the display options :

- Grid.
- Units.
- Cursor.
- Invisible pins.
- Allowed directions of wires and buses.

## 2.11 - Pop-up menus and quick editing

A right click opens a pop-up menu which content depends on the element selected, if any. You have immediate access to:

- Zoom factor.
- Grid adjustment.
- And according to the case, editing of the most usually modified parameters.

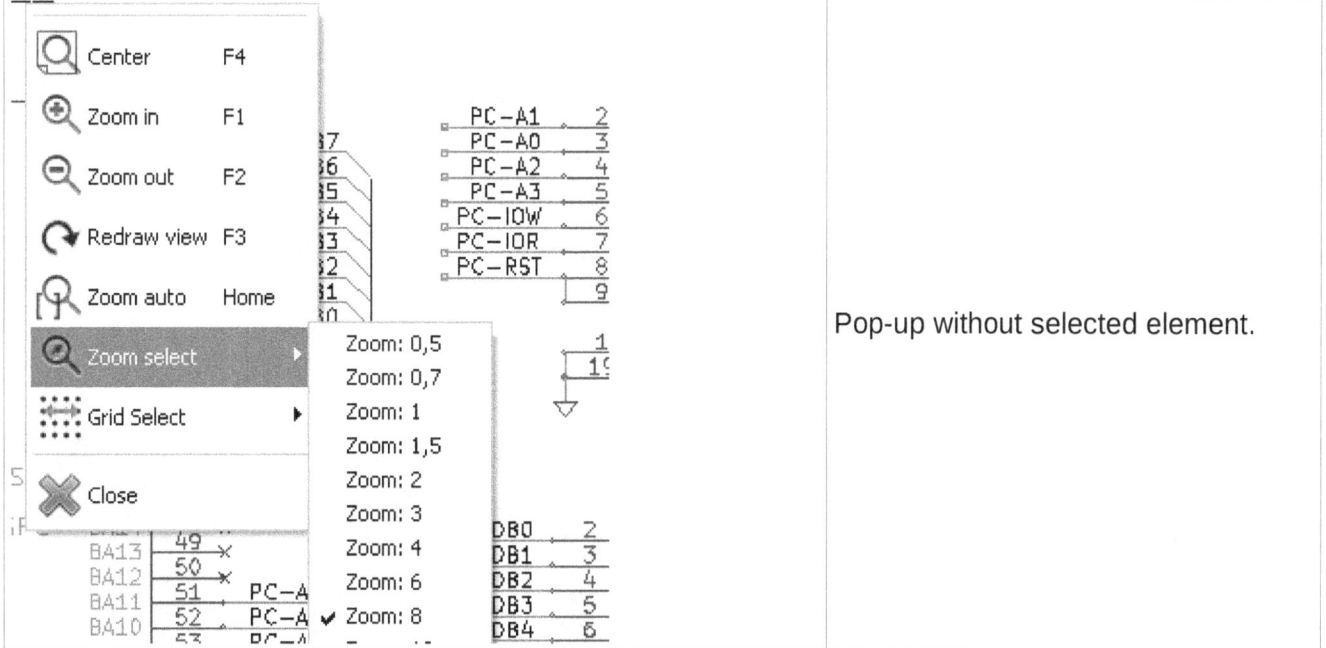

Pop-up without selected element.

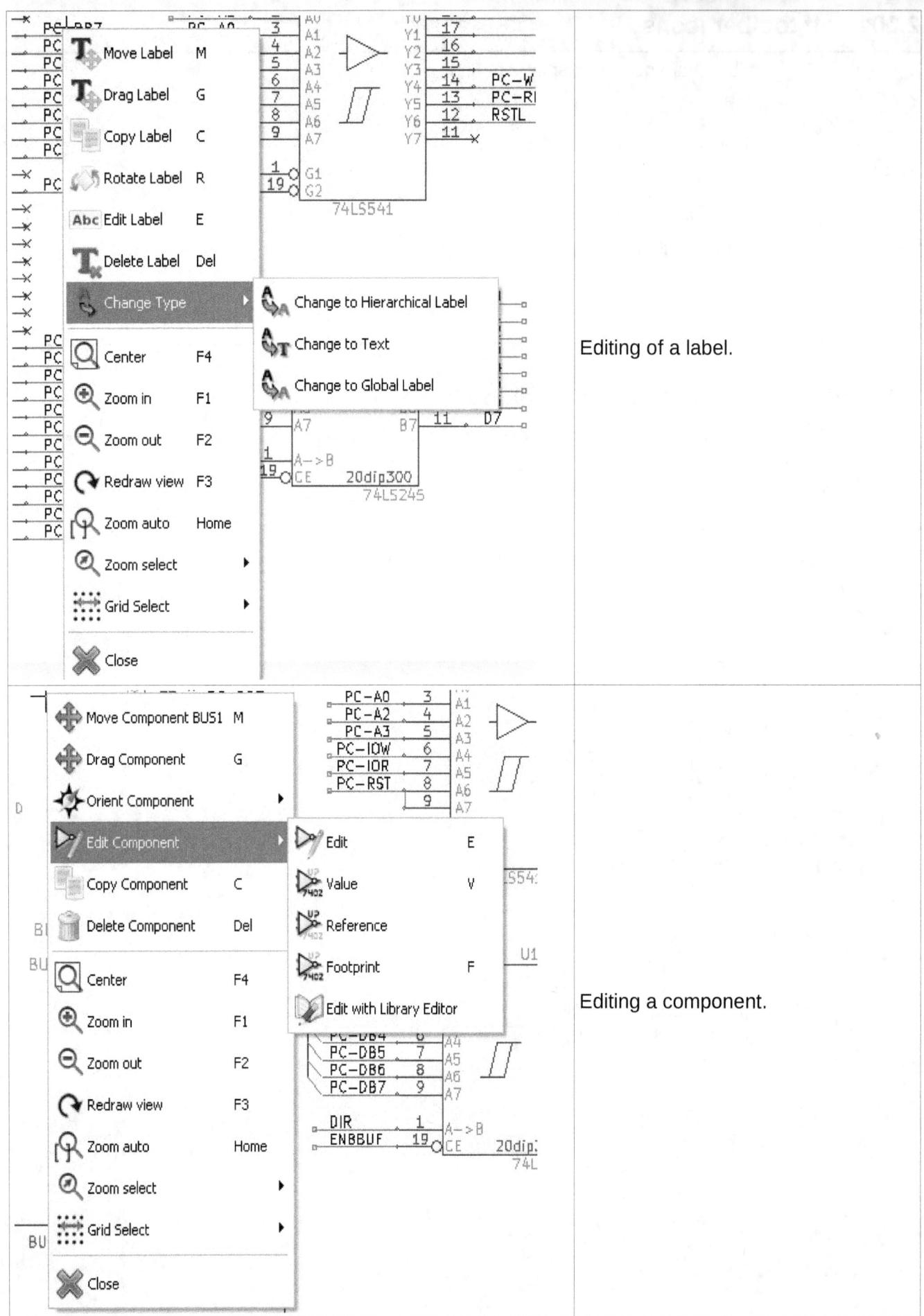

Editing of a label.

Editing a component.

# 3 - Main top menu

## Table of Contents

## 3.1 - File menu

Here you can see what the "File" menu looks like.

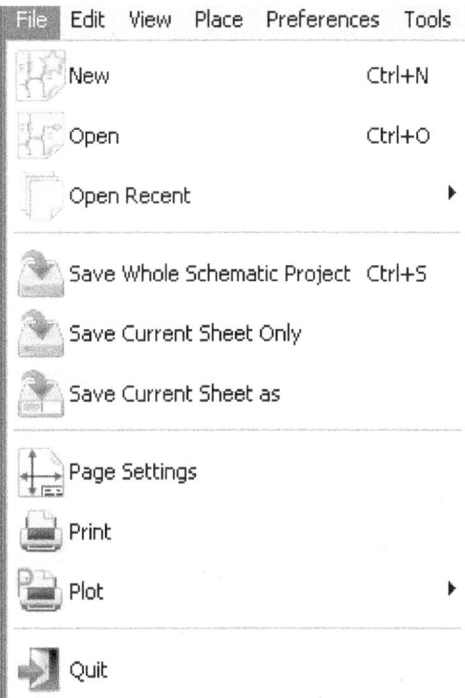

| New | Clear current schematic and initialize a new one |
|---|---|
| Open | Load a  schematic hierarchy |
| Open Recent | Open a list of recent opened files for loading |
| Save Whole Schematic project | Save current sheet and all its hierarchy. |
| Save Current Sheet Only | Save current sheet, but not others in a hierarchy. |
| Save Current sheet as... | Save current sheet with a new name. |

| Print | Access to print menu (See also chap "Print and Plot"). |
|---|---|
| Plot | Plot in Postscript HPGL or SVF format (See chap "Print and Plot"). |
| Quit | Quit without saving. |

## 3.2 - Preferences menu

### 3.2.1 - Preferences

| Library | Select libraries and the library's path |
|---|---|
| Colors | Select colors. |
| Options | Display options (Units, Grid size.). |
| Language | Access to the current list of translations. Use default. Mainly for translators and developers |
| Read  preferences Save preferences | Read and Save configuration file. |
| Hotkeys | Access to the hot keys menu |

### 3.2.2 - Hot-keys sub menu

| List Current Keys | Shows the current hotkeys, Same as the hotkey "?" |
|---|---|
| Edit Hokeys | Launch the hotkeys editor |
| Export Hotkeys Config | Create a hotkeys configure file. |
| Import Hotkeys Config | Read a previously exported hotkeys configure file. |

### 3.2.3 - Preferences menu / Libs and Dir

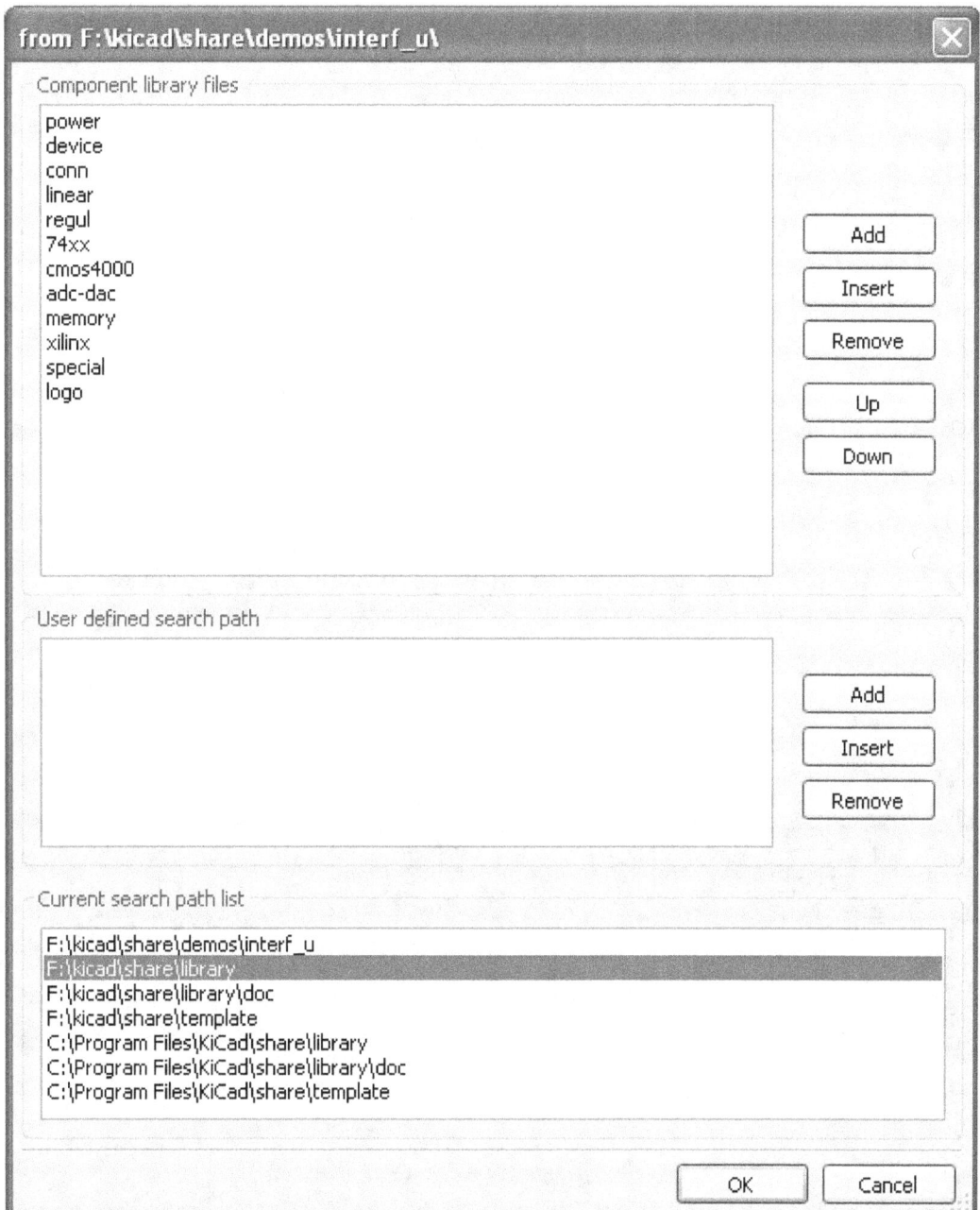

Eeschema configuration is essentially used for:

- Library's path.
- Library's list.

The configuration parameters are saved in the .pro file. Different configuration files in different directories are also possible.

Eeschema seeks and uses by decreasing priorities:

1. The configuration file (project>.pro) in the current directory.
2. The kicad.pro configuration file in the kicad directory. This file can thus be the default configuration.
3. Default values if no file is found. It will at least then be necessary to fill out the list of libraries to load, and then save the configuration.

### 3.2.4 - Preferences menu and colors

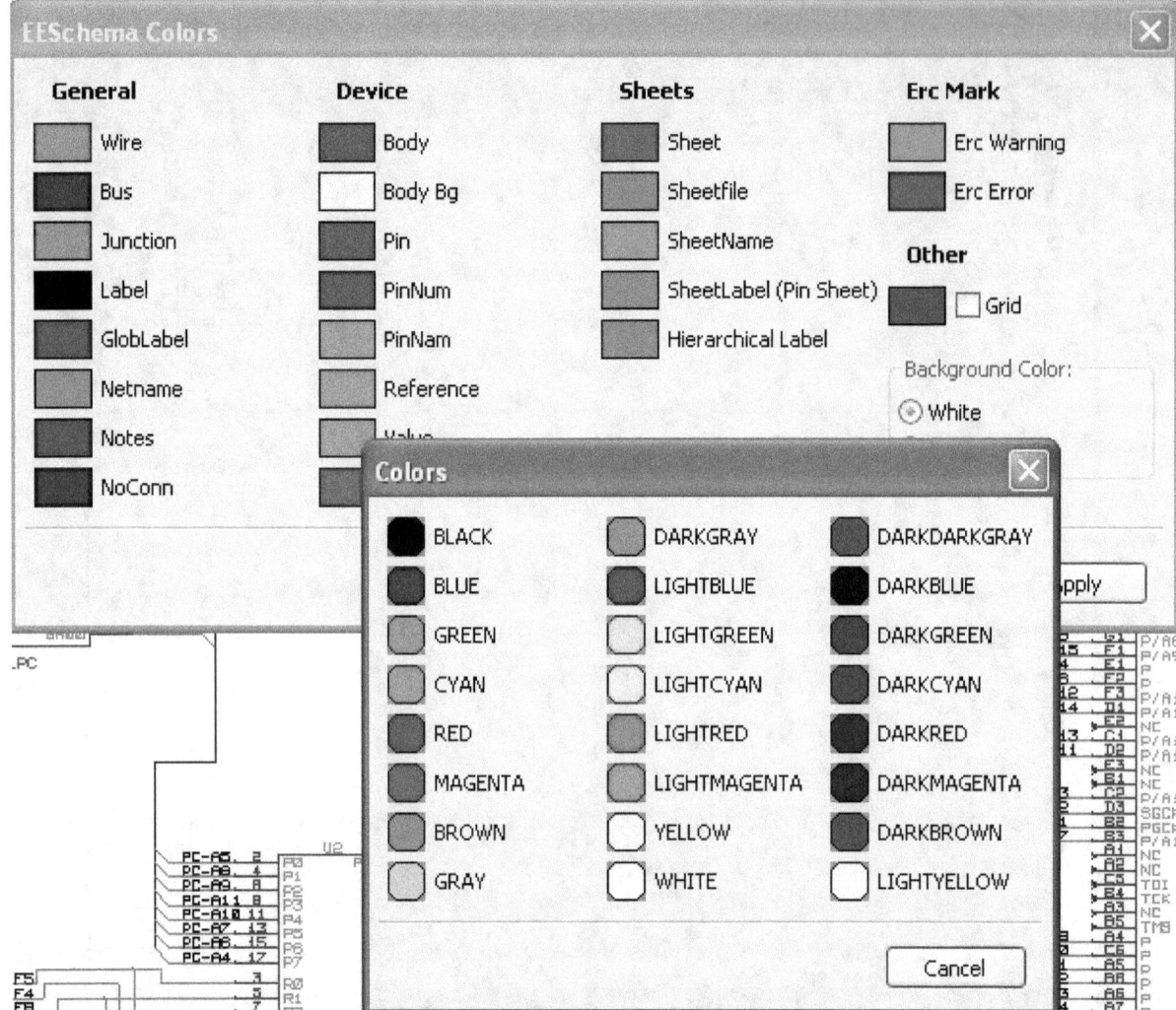

Various drawing elements, colur selection and background colur (black or white only).

## 3.2.5 - Preferences and Options

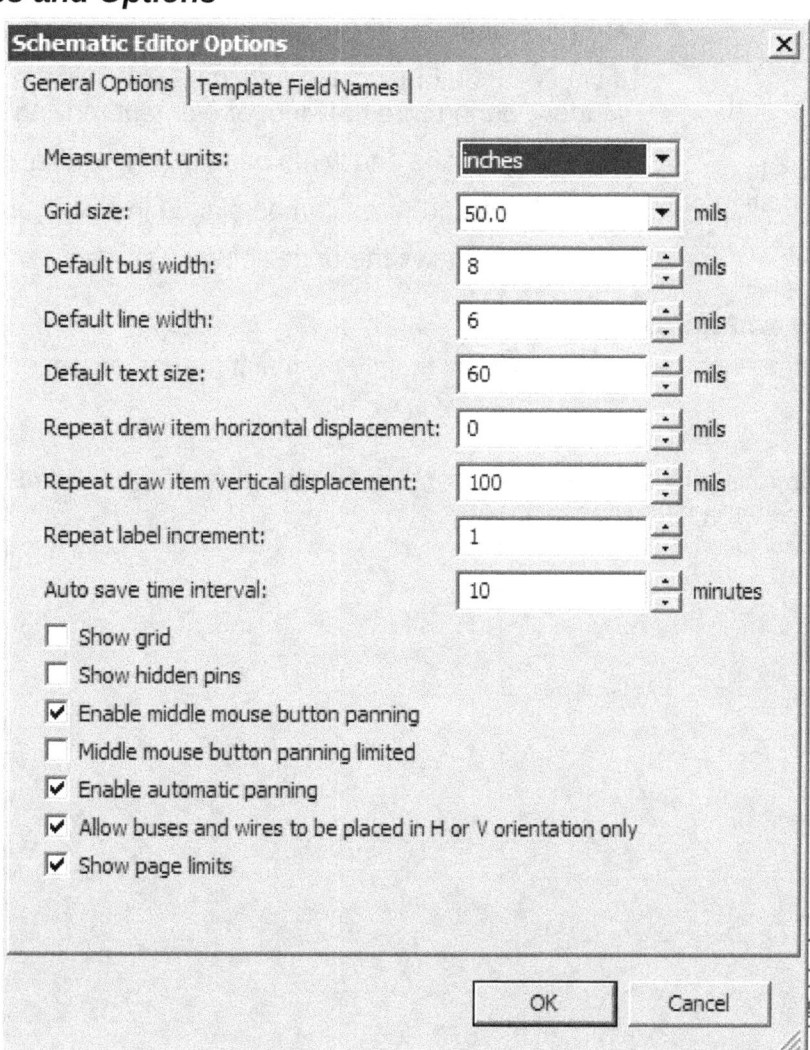

| Measurement units: | Select the display and the cursor coordinate units (Inches or Millimeters). |
|---|---|
| Grid Size: | Grid size selection.<br>**One must work with normal grid (0,050 inches or 1,27 mm).** <u>Smaller grids are used for component building</u>. |
| Default line width: | Pen size used to draw objects that do not have a specified pen size. |
| Default text size: | Value used when creating new texts or labels |
| Repeat draw item horizontal displacement | shift value on X axis during element duplication (usual value 0)<br>(after placing an item like a component, label or wire, a duplication is made by the *Inser* key) |
| Repeat draw item horizontal displacement | shift value on Y axis during element duplication (usual value is 0,100 inches or 2,54 mm) |
| Repeat label increment: | Increment during duplication of texts ending in a number, such as bus members (usual value 1 or - 1). |
| Show Grid : | If checked : display grid. |
| Show hidden pins: | Display invisible (or *hidden*) pins .<br>If checked, allows the display of power pins. |
| Enable middle mouse button panning | When enabled, when the middle mouse button is pressed, the entire sheet is moved, following the cursor |

| Middle mouse button panning limited | When enabled, the middle mouse button cannot move the sheet area "outside" the displayed area. |
|---|---|
| Enable automatic panning | If checked, automatically shifts the window if the cursor leaves the window, during wire drawing, or element moving. |
| Allow buses and wires to be placed in H or V orientation only | If checked buses and wires can be only vertical or horizontal. Else buses and wires can be placed in any direction. |
| Show page limit | If checked, shows the page limits on screen. |

### 3.2.6 - Preferences and Language

Use default mode. Other languages are available mainly for maintenance purpose.

## 3.3 - Help menu

Access to on-line help (this document) for an extensive tutorial about KiCad and also for checking the current version of Eeschema (Eeschema about).

## Table of Contents

## 4.1 - Sheet Management

With the icon you have access to the sheet settings. Here, you can define the sheet size and various text sections in the title block on the bottom right-hand corner.

**Page Settings**

| Page Size: | |
|---|---|
| ⦿ Size A4 | |
| ○ Size A3 | |
| ○ Size A2 | |
| ○ Size A1 | |
| ○ Size A0 | |
| ○ Size A | |
| ○ Size B | |
| ○ Size C | |
| ○ Size D | |
| ○ Size E | |
| ○ User size | |

Number of sheets: 1     Sheet number: 1

Revision:
[          ]    ☐ Export to other sheets

Title:
[                                    ]    ☐ Export to other sheets

Company:
[                                    ]    ☐ Export to other sheets

Comment1:
[                                    ]    ☐ Export to other sheets

Comment2:
[                                    ]    ☐ Export to other sheets

Comment3:
[                                    ]    ☐ Export to other sheets

Comment4:
[                                    ]    ☐ Export to other sheets

User Page Size X:
17,000

User Page Size Y:
11,000

OK     Cancel

The date is automatically updated. Total number of sheets and sheet number are automatically updated.

## 4.2 - Options of the schematic editor

### 4.2.1 - General options

These options are relative to the drawings

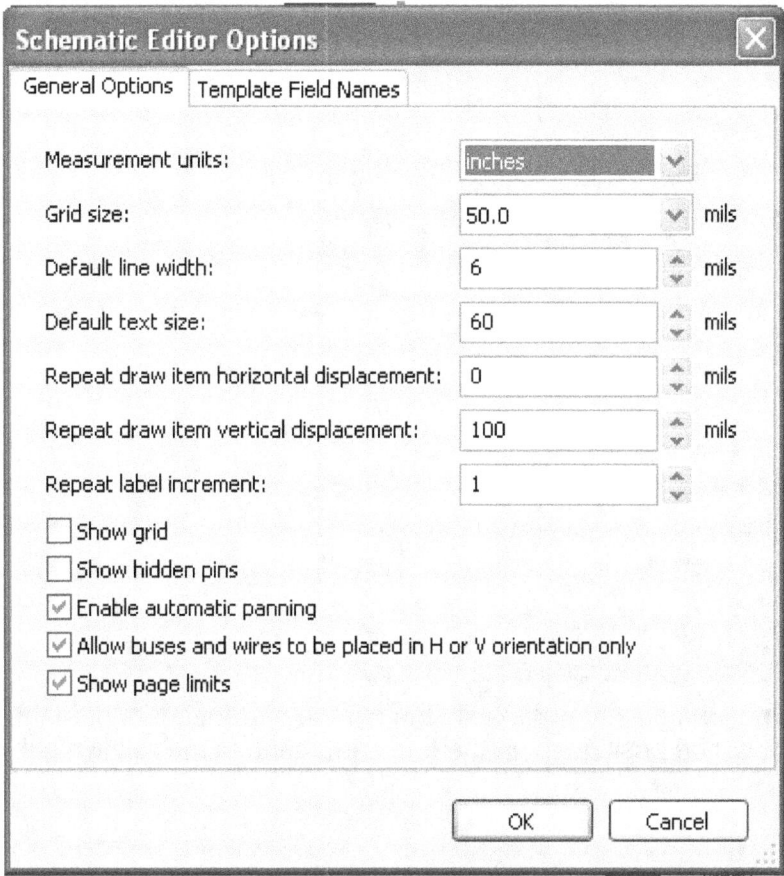

## 4.2.2 - Template fields names

You can define custom fields that will always existing in each component (even if the fields are left empty in a given component).

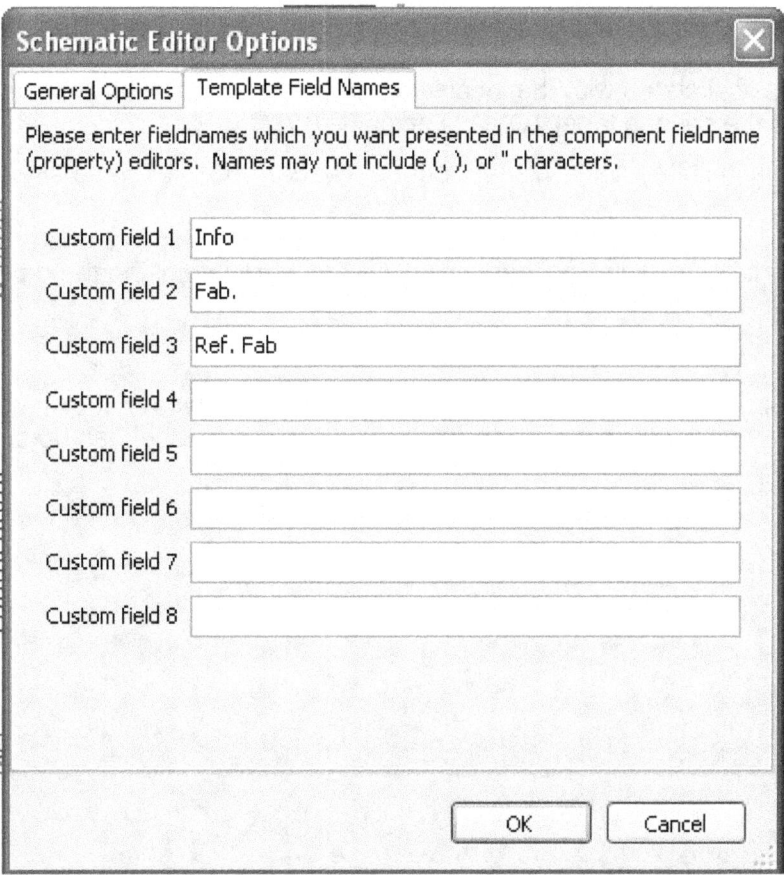

## 4.3 - Search tool

the following icon 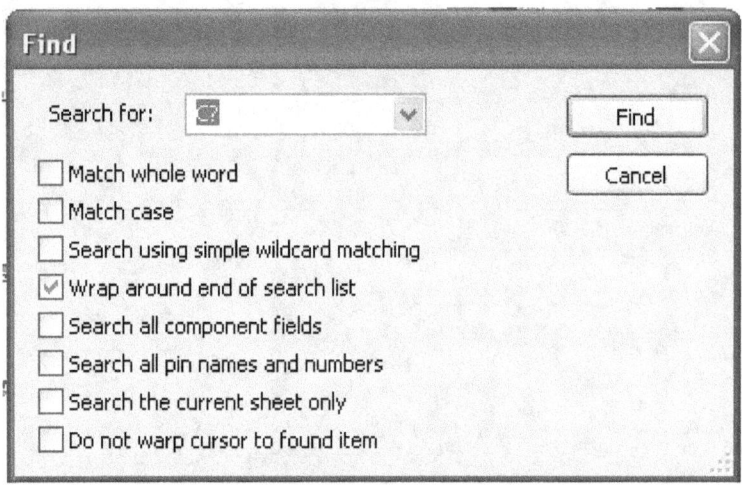 can be used to access the search tool.

You can search a component, a value, or a text string in the current sheet or in the whole hierarchy. Once found, the cursor will be positioned on the found element, in the relative sub-sheet.

## 4.4 - Netlist tool

The icon ![.net] gives access to the netlist tool used to generate a netlist file.

This netlist file can apply to the whole hierarchy (usual option), or only to the current sheet (the netlist is then partial, but this option can be useful for some software).

In a multisheet hierarchy, any local label is known only inside the sheet to which it belongs.

Thus the label TOTO of sheet 3 is different from the label TOTO of sheet 5 (if no connection has been intentionally introduced to connect them). This is due to the fact that the sheet number (updated by the annotate command) is associated with the local label. In the previous example, the first label TOTO is actually TOTO_3, and the second label TOTO is actually TOTO_5.

This association can be inhibited if it is wished, but be aware of possible undesired connections.

Note 1:

Label lengths have no limitations in Eeschema, but the software exploiting the generated netlist can be limited on this point.

Note 2:

Avoid spaces in the labels, because they will appear as separated words. It is not a limitation of Eeschema, but of many netlist formats, which often suppose that a label has no spaces.

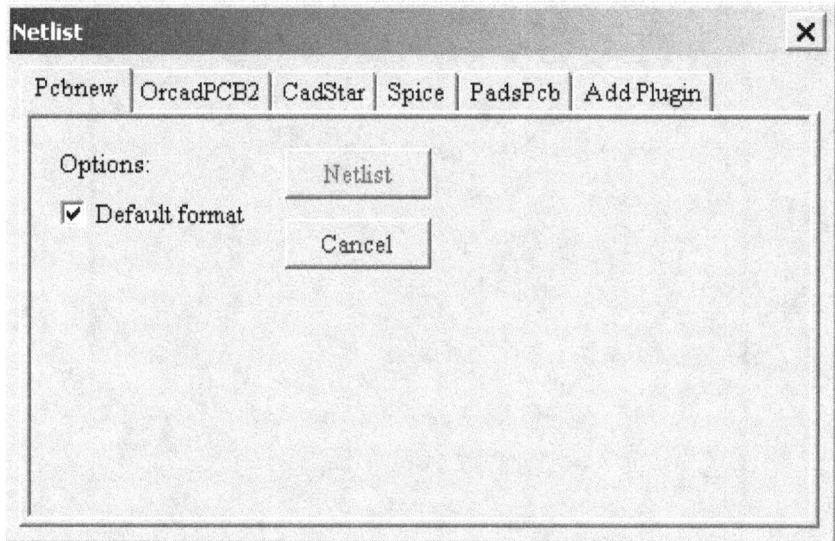

Option:

Default Format :

Check to select Pcbnew as the default format.

Other formats can also be generated :

- Orcad PCB2
- CadStar
- Spice, for the Spice simulator.

External plugins can be launch to extend the netlist formats list (a PadsPcb Plugin was added here)

## 4.5 - Annotation tool

The icon 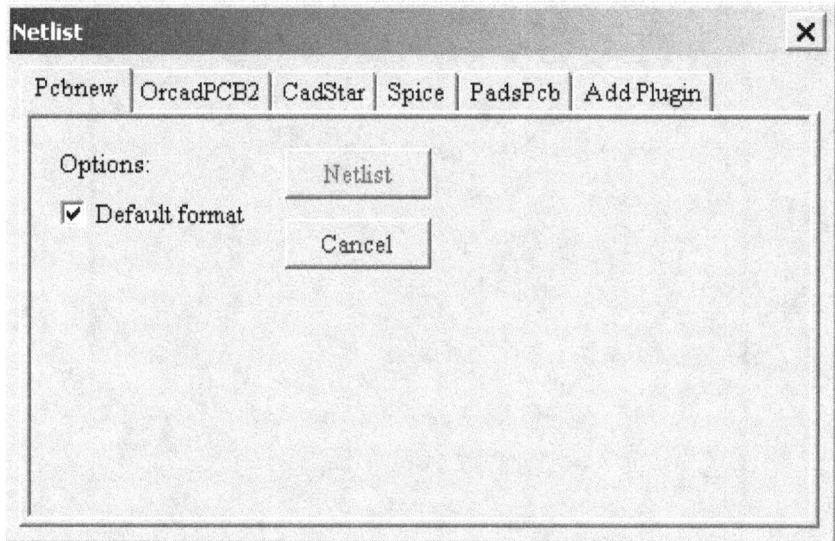 gives access to the annotation tool. This tool performs an automatic naming  for all used components.

For multi-part components (such as 7400 TTL which contains 4 gates), a multi-part suffix is also allocated (thus a 7400 TTL designated U3 will be divided into U3A, U3B, U3C and U3D).

You can unconditionally annotate all the components, or only the new components, i.e. those which were not previously annotated.

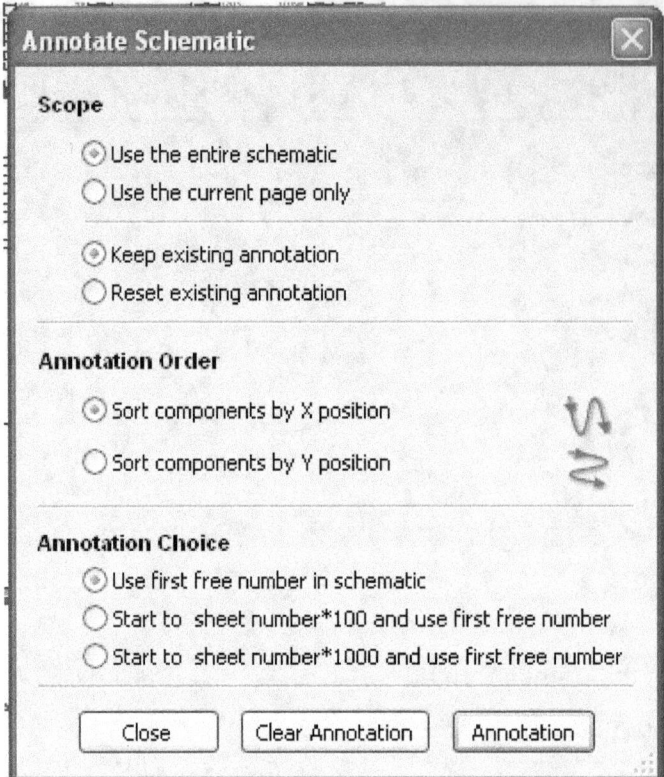

**Scope**

1) Use the entire schematic. All the sheets are re-annotated (usual Option).

2) Use the current page only. Only the current sheet is re-annotated (this option is to be used only in special cases, for example to evaluate the amount of resistors in the current sheet.).

3) Keep existing annotation. Conditional annotation, only the new components will be re-annotated (usual option).

4) Reset existing annotation. Unconditional annotation, all the components will be re-annotated (this option is to be used when there are duplicated references).

**Order**

Sorting option to set the annotation numbers to components

## 4.6 - Electrical Rules Check tool

The icon ![icon] gives access to the electrical rules check (ERC) tool.

This tool performs a design verification and is particularly useful to detect forgotten connections, and inconsistencies.

Once you have run ERC, Eeschema places markers on the pins or labels able to highlight a problem. The diagnosis can then be given by left clicking on the marker. An errors file can also be generated.

### 4.6.1 - Main ERC dialog

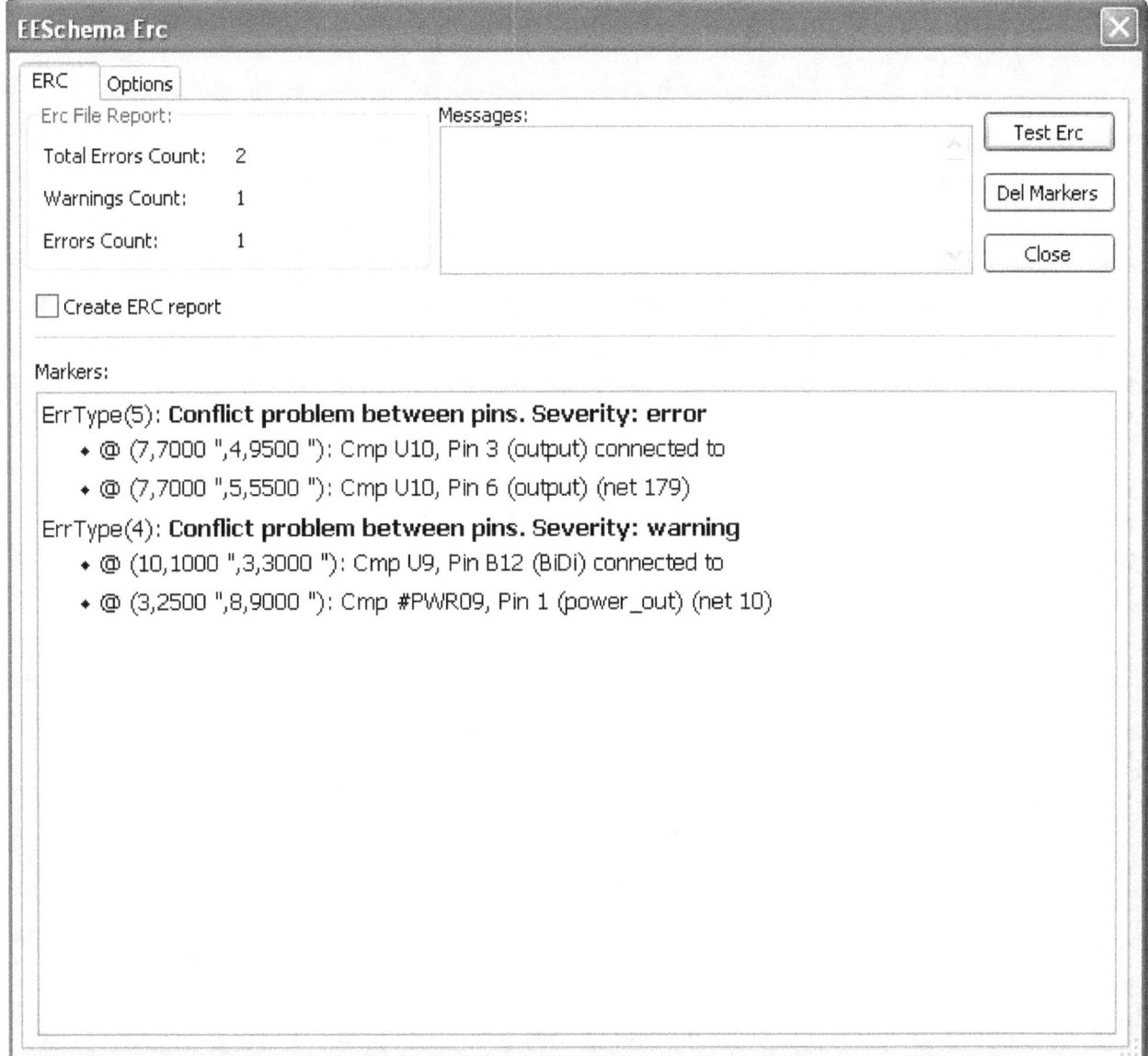

Errors are displayed in the Erc Diags dialog box:

- Errors and warnings count.
- Errors count.
- Warnings count.

Option:

- Create the ERC report: check this option to generate an ERC report file.

Commands:

- Test Erc: to perform an Electrical Rules Check.
- Del Markers: to remove all ERC markers.
- Close: to exit this dialog box.

Note:

- When clicking on an error message, jump to the corresponding marker in schematic.

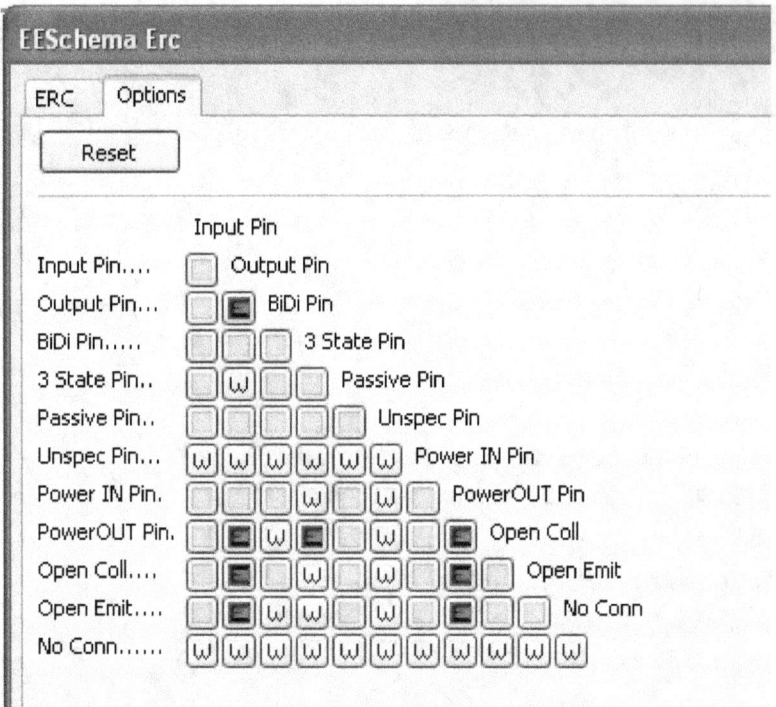

This Setup ERC dialog box allows you to establish connectivity rules between pins; you can choose between 3 options for each case:

- No error
- Warning
- Error

Each square of the matrix can be modified by clicking on it.

## 4.7 - Bill of Material tool

The icon ![BOM] gives access to the bill of material (BOM). This menu allows the generation of a file listing of the components and/or hierarchical connections (global labels).

Components can be sorted by:

- Reference.
- Value.

And multi-part components can be detailed. Global labels can be sorted by :

- Alphabetical classification
- Sub-sheet

Different kinds of sorting can be used simultaneously. Options are:

| Components by Reference | Bill of Material sorted by Reference. |
|---|---|
| Component by Value | Bill of Material sorted by Value. |
| Sub components | The BOM shows every device of multi-part components (ex U2A, U2B…). |
| Hierarchy Pins by name | Hierarchical connections sorted alphabetically. |
| Hierarchy Pins by Sheet | Hierarchical connections sorted by sheet number. |
| List | Creates a plain text file ready to print |
| Text for spreadsheet import | Creates an ASCII file which can be easily imported in a **spreadsheet** |
| Single Part per line | Creates a csv file combining components with the same Value into a single line, listing reference designators comma separated. |
| Launch list browser | Run the text editor to load and display the BOM list file after creating. |

A useful set of component properties to use for a BOM are:

- Value – unique name for each part used.
- Footprint – either manually entered or back-annotated (see below).
- Field1 – Manufacturer's name.
- Field2 – Manufacturer's Part Number.
- Field3 – Distributor's Part Number.

For example:

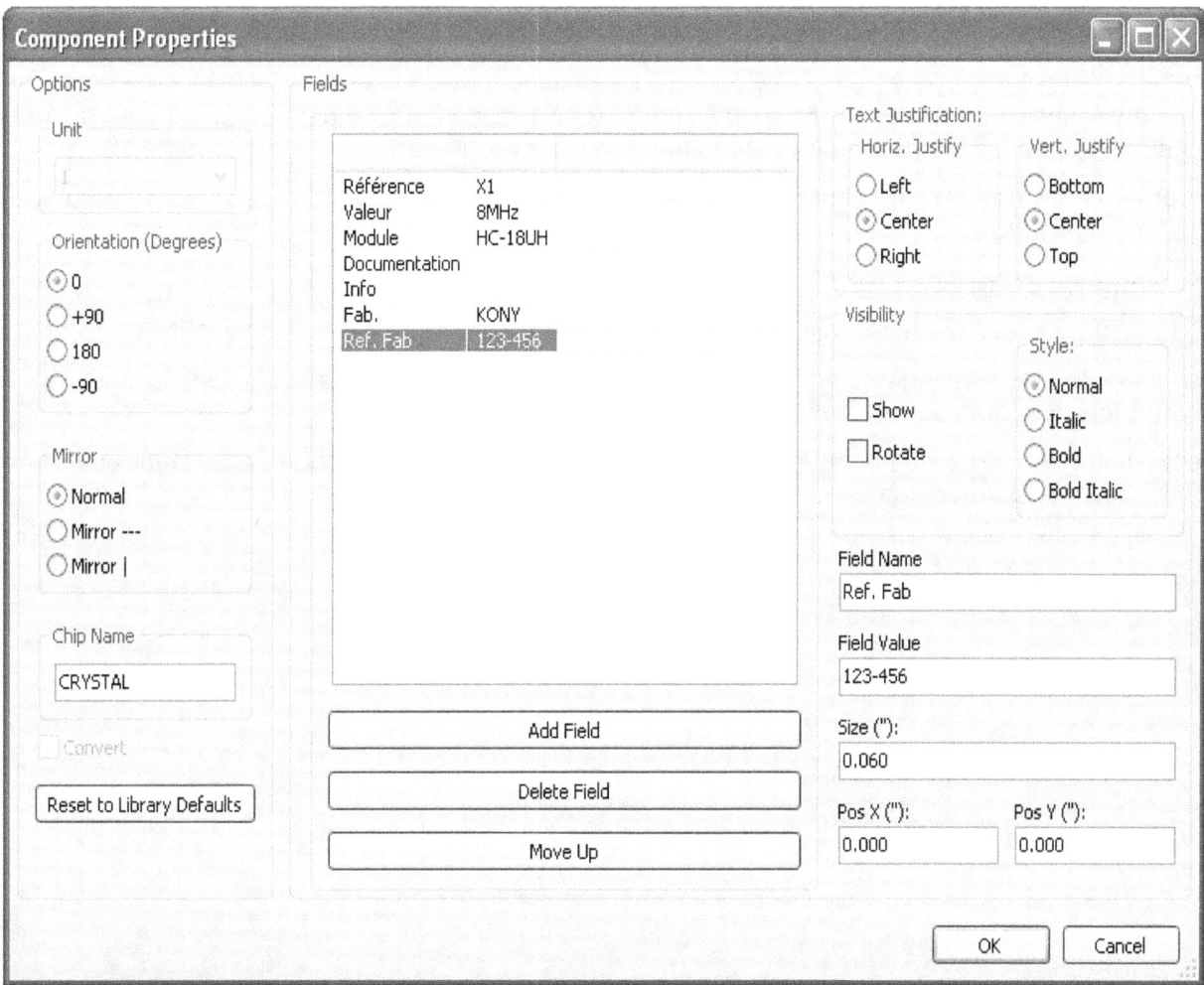

Using the BOM Format Single Part per line only requires the component properties to be edited for one component on the schematic and not all components with that same Value.

However, if there are different parts, both with a Value of 33K, may be one is 1/10 W and another is ¼ W, or may have a different footprint, specify one as 33K and the other as 33KBig and these will be listed as different parts.

The output is in a format than can be imported into a spreadsheet where cost numbers (or optionally even Field4) may be added to derive a board cost and assist with parts procurement.

# 4.8 - Import tool for footprint assignment:

## 4.8.1 - Access:

The icon ![BACK icon] gives access to the back-annotate tool.

This tool allows a schematic to be captured, make footprint assignments using Cvpcb's table and browser tools, then export that assignment back to the schematic.

This function reads the .cmp file previously created by Cvpcb and initialize the footprint field (Field 3) of components.

This is not mandatory for Pcbnew, but useful to add the footprint field when creating the Bill of Material and the netlist.

This feature keeps the component footprint/reference information in a single source file, the schematic, which is the source for the netlist and makes the .cmp file redundant.

The footprint assignments will appear in any future netlist export from Eeschema. This is useful when using some netlist formats.

## 4.8.2 - Note for Pcbnew

Using the .cmp file or the netlist only to assign a footprint to a component is a choice inside Pcbnew.

When Pcbnew does not find a .cmp file corresponding to the .net file, it uses the component footprint/reference found in the .net file.

However, using the .cmp file is better, because if the designer changes a footprint assignment from Pcbnew, the corresponding .cmp file is also updated.

# 5 - Schematic Creation and Editing

## Table of Contents

## 5.1 - Introduction

A schematic can be represented on a single sheet, but, if big enough, it will require several sheets.

A schematic represented on several sheets is then called hierarchical, and all its sheets (each one represented by its own file) constitutes a Eeschema project. A project consists of a main schematic, called the root schematic, and sub-sheets constituting a hierarchy.

In order to find every file of the project, you will have to follow drawing rules which will be described hereafter.

In the following, when we talk about project, we will be referring to both single sheet and hierarchical multi sheets. An additional special chapter explains the use of the hierarchy and its characteristics.

## 5.2 - General considerations

A schematic designed with Eeschema is more than a simple graphic representation of an electronic device. It is normally the entry point of a development chain which allows for:

- The control of the electrical rules (ERC) that allows the detection of errors or omissions in the schematic.
- The automatic generation of the bill of material (BOM).
- The netlist generation for simulation software such as Pspice.

- The netlist generation for the generation of a printed circuits board design using PcbNew. The consistency check between the schematic and the printed circuit board is then automatic and instantaneous.

In order to benefit from all these options, you will have to respect certain constraints and conventions which will also enable you to avoid nasty surprises.

A schematic mainly consists of components, wires, labels, junctions, buses and power ports. For clearness in the schematic, you can place purely graphical elements like bus entries, comments, and dotted lines to draw frames.

## 5.3 - The development chain

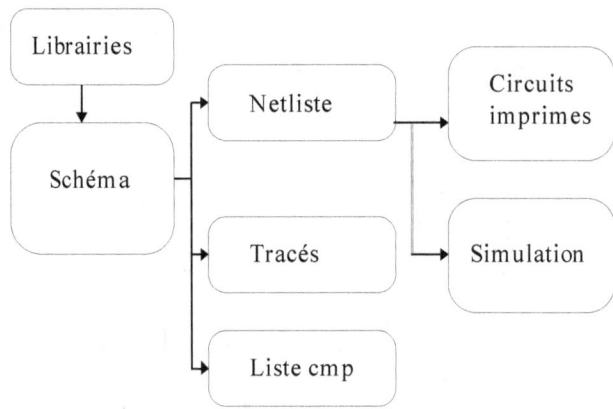

The schematic software uses component libraries. In addition to the schematic design file, the netlist file is particularly important because it is used by the other design software.

A netlist file gives the list of the components and connections resulting from the schematic.

There is (unfortunately for the user) a great number of netlist formats, some are more popular then others. It is the case of the Spice format for example.

## 5.4 - Component placement and editing

### 5.4.1 - Find and place a component

To load a component in your schematic you can use the icon ![icon]. To place a new component, click at the place you want to draw it. A dialog box allows you to type the name of the module to load.

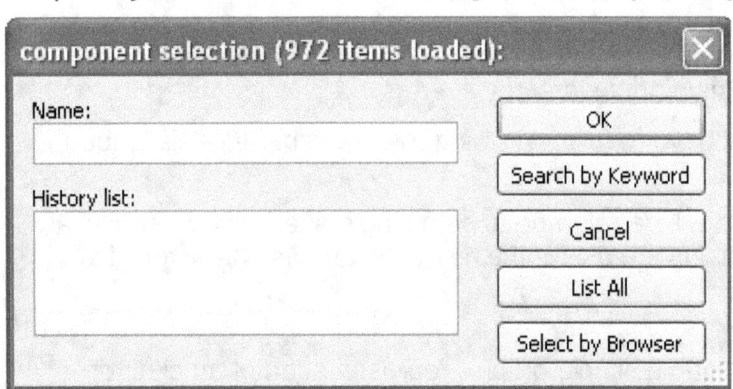

The dialog box displays the last two elements loaded.

If you type *, or if you select the button "list all", Eeschema will display the libraries list, and then the available components.

If you type the symbol "=" followed by key words, EESchema will then display a list of components according to all the key words.

You can also list a selection. For example if you enter LM2 *, all the component's names starting with LM2 will be listed

The selected component will appear on the screen, in placement mode.

Before the component gets placed in the desired position (with a left click), you can rotate the component by 90 degrees, have a mirror view according to axis X or Y, or select its representation via the fast edit pop-up menu. This can also easily be done after placement.

If the desired component does not exist, remember that you can often load a similar component and modify it : if a 54LS00 is wanted, you can obviously load a 74LS00 and change the name 74LS00 to 54LS00.

Here is a component during placement:

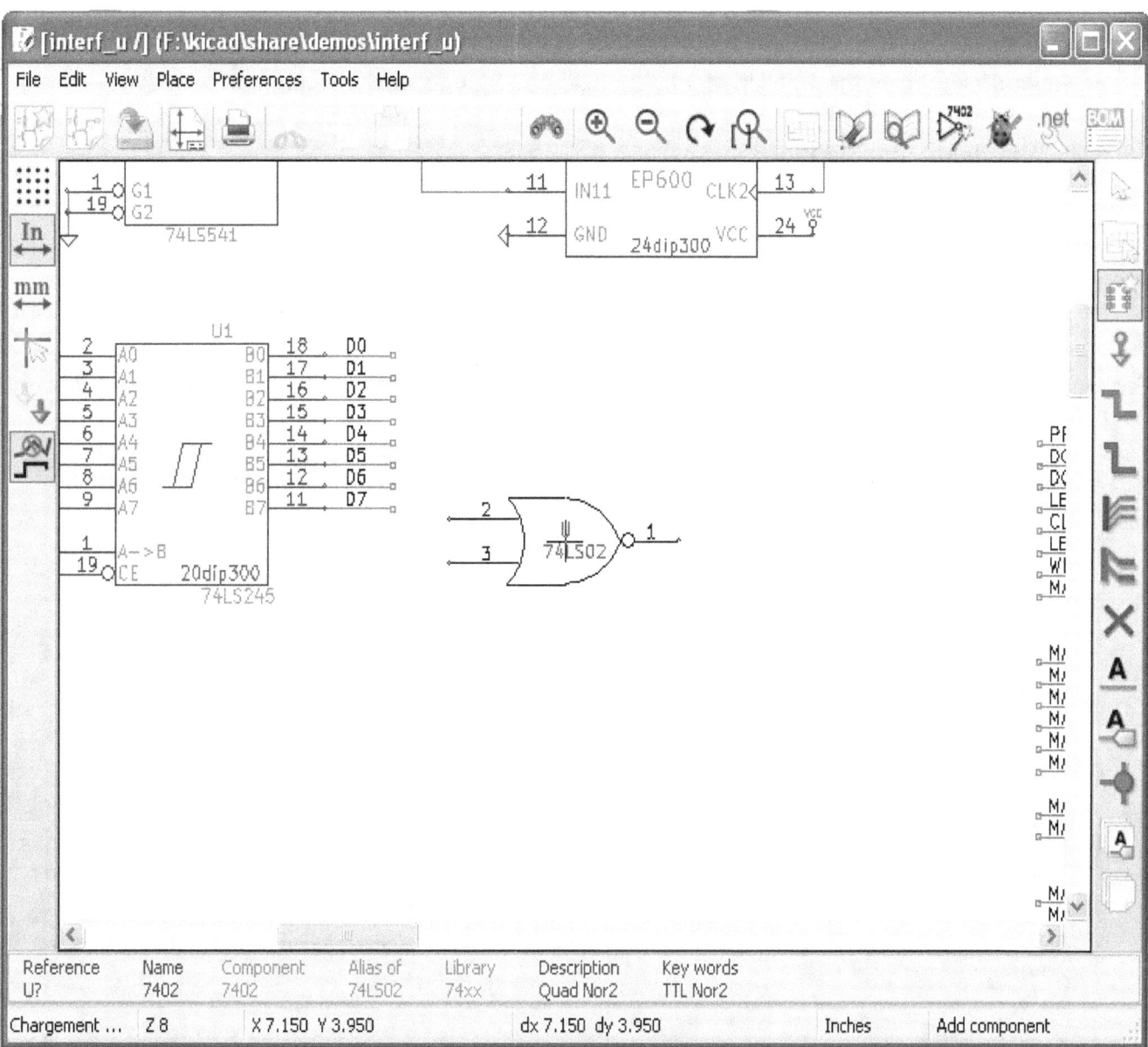

### 5.4.2 - Power ports

A power port symbol is a component (the symbols are grouped in the "power" library). So you can use the previous command. But as these placements are frequent, the ⚓ tool is available. This tool is similar to the preceding one, except that the search is done directly in the "power library", saving time.

### 5.4.3 - Component Editing and Modification (already placed component)

The editing and modification of a component can be of two types

- Modification of the component itself (position, orientation, part selection of a multi-part component).
- Modification of one of the fields (reference, value, or others) of the component.

When a component has just been placed, you may have to modify its value (particularly for resistors, capacitors, etc.), but it is useless to assign to it a reference number right away, or to select the part of a multi-part component (like a 7400).

This can be done automatically by the annotation function.

### 5.4.3.1 - Component modification

To modify some feature of a component, position the cursor mouse on the component (not to position on a field). One can then:

- Double-click on the component to open the full editing dialog box.
- Right-click to open the Pop Up menu, and use one of the displayed commands (Move, Orientation, Edit, Delete).

### 5.4.3.2 - Text fields modification

You can modify the reference, value, position, orientation, size and the visibility of the fields. For simple editing:

- Double-click on the text field to modify it.
- Right-click and use one of the displayed commands (Move, Rotate, Edit, Delete) in the Pop Up menu.

For a more complete editing option, or in order to create fields, double-click on the component. This will open the "component properties" dialog box.

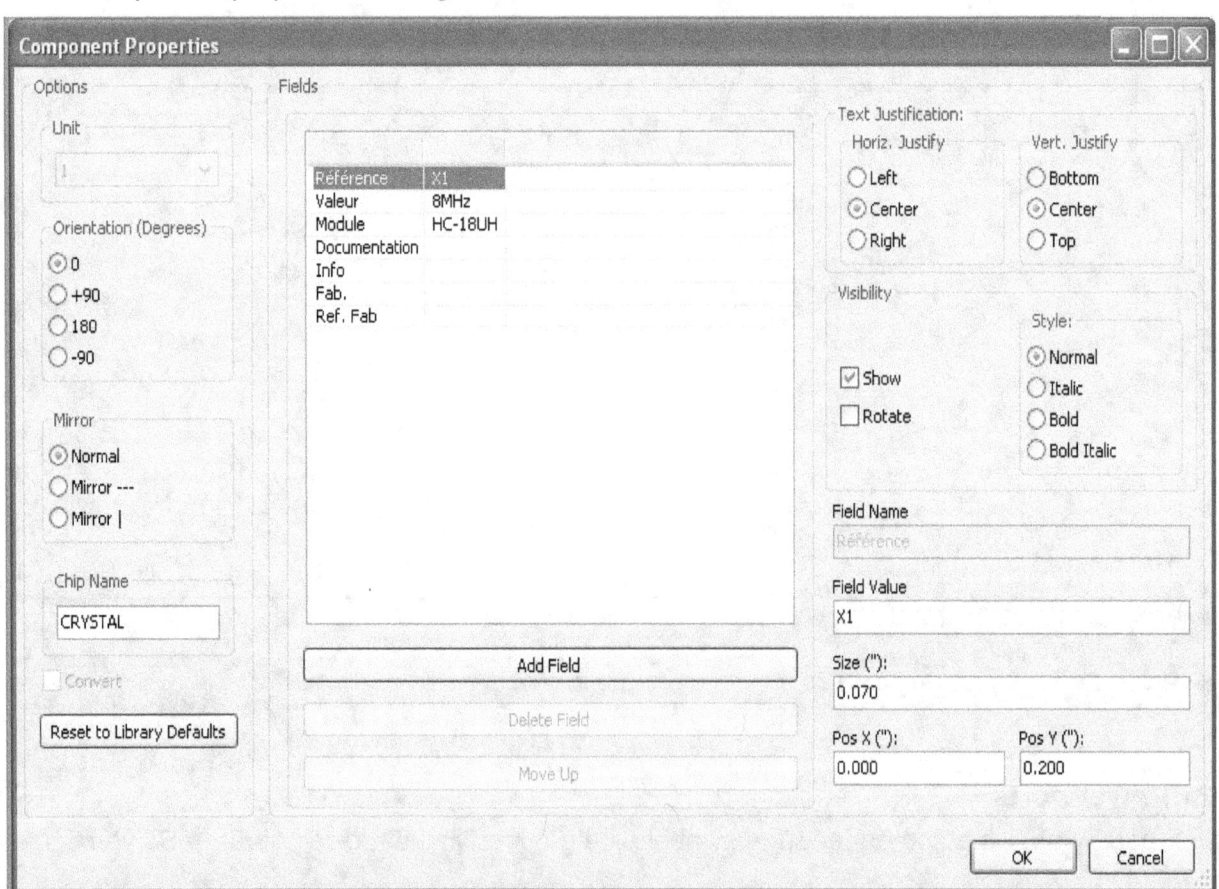

You can set the orientation and others options of the component, and edit, add or remove fields.

Each field can be visible or not, and displayed horizontally or vertically. The displayed (and changeable) position is always indicated for a normally displayed component (no rotation or mirror) and relates to the anchoring point of the component.

The option "Reset to Library Defaults" set the component to the orientation 0, and the options, size and position of each field. However, texts fields are not modified because this could break the schematic.

## 5.5 - Wires, Buses, Labels, Power ports

### 5.5.1 - Introduction

All these drawing elements can also be placed with the tools on the vertical right toolbar.

These elements are:

- **Wires.** Typical  usual connections.
- **Buses.** To connect bus labels, for esthetic considerations of the drawing.
- **Dotted lines**. For graphic presentation.
- **Junctions.** To force connections between crossing wires or buses.
- **Bus entries** of Wire to Bus or Bus to Bus connections. For aesthetic considerations of the drawing.
- **Labels.** For usual connections.
- **Global labels**. For connections between sheets.
- **Texts.** For commenting.
- "**No Connection**" symbols. To end a pin that does not need any connection.
- **Hierarchy sheets**, and their connection pins.

### 5.5.2 - Connections (Wires and Labels)

There are two ways to establish connection:

- Pin to pin wires.
- Labels.

The following figure shows the two methods:

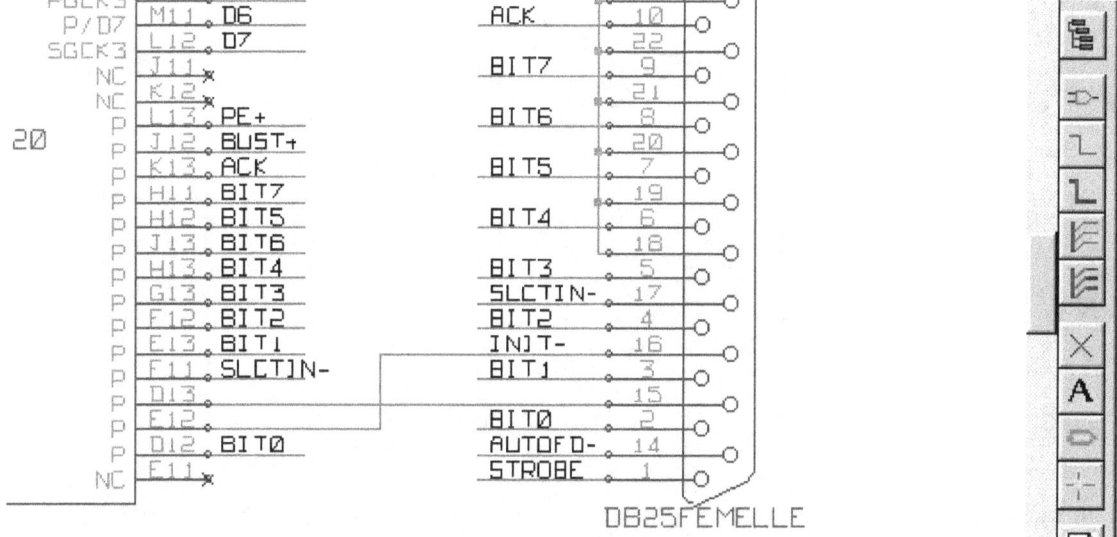

*Note 1:*

The point of "contact" (or anchoring) of a label is the lower left corner of the first letter of the label.

This point must thus be in contact with the wire, or be superimposed at the point of contact of a pin so that this label is taken into account.

*Note 2:*

To establish a connection, a segment of wire must be connected by its ends to an another segment or to a pin.

If there is overlapping (if a wire passes over a pin, but without being connected to the pin end) there is no connection. However, a label will be connected to a wire whatever the position of the anchoring point of the label on this wire.

*Note 3:*

If a wire must be connected to another wire, otherwise than by their ends, it will be necessary to place a junction symbol at the crossing point.

The previous figure (wires connected to DB25FEMALE pins 22, 21, 20, 19) shows such a case of connection using a junction symbol.

*Note 4:*

If two different labels are placed on the same wire, they are connected together and become equivalent: all the other elements connected to one or the other labels are then connected to all of them.

### 5.5.3 - Connections (Buses)

Let us consider the following schematic:

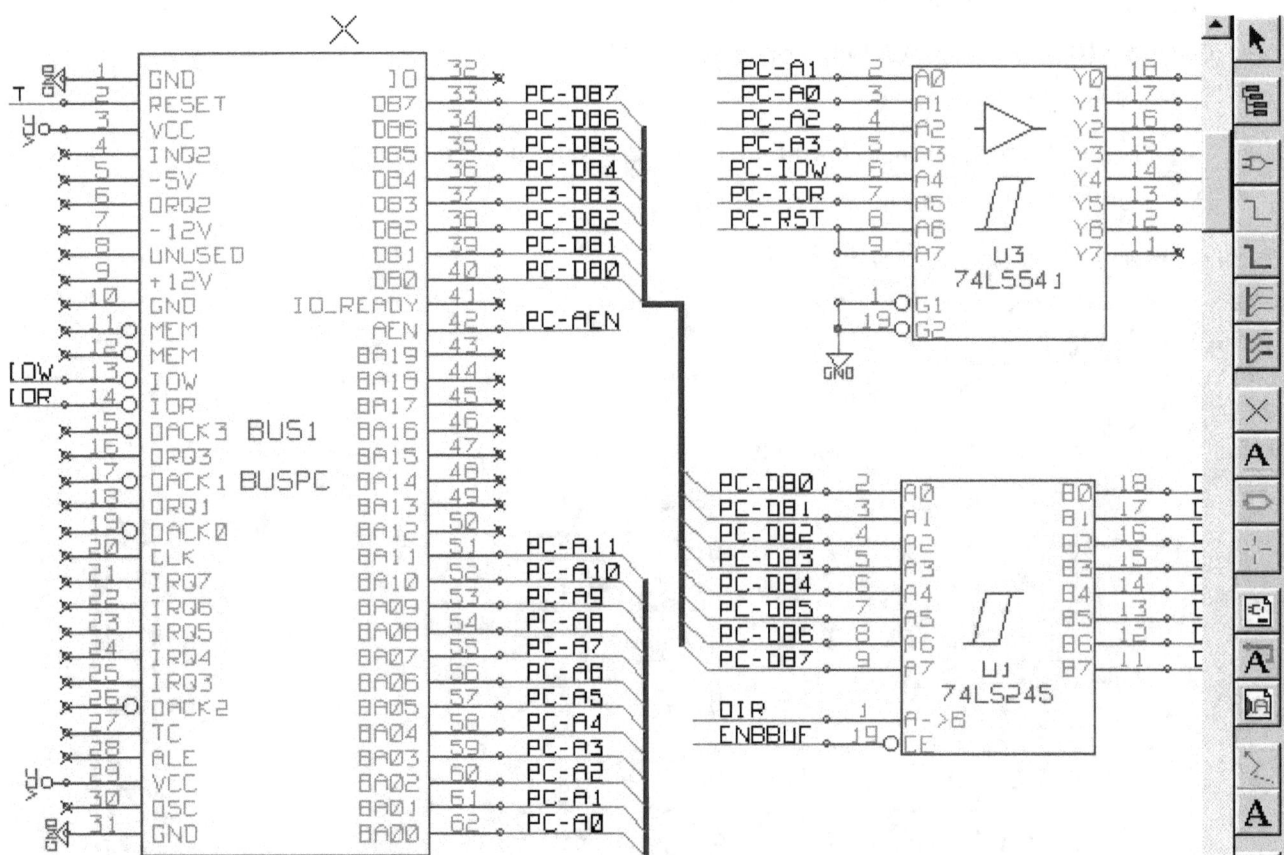

Many pins (particularly component U1 and BUS1) are connected to buses.

### 5.5.3.1 - Bus members

From the schematic point of view, a bus is a collection of signals, starting with a common prefix, and ending by a number. This concept is not exactly the one which is used for a microprocessor bus. Each signal is a member of the bus. PCA0, PCA1, PCA2, are thus members of PCA bus.

The complete bus is named PCA [N. .m], where N and m are the first and the last wire number of this bus. Thus if PCA has 20 members from 0 to 19, the complete bus is noted PCA [0..19]. But a collection of signals like PCA0, PCA1, PCA2, WRITE, READ cannot be contained in a bus.

### 5.5.3.2 - Connections between bus members

Pins connected between the same members of a bus must be connected by labels. Indeed, directly connecting a pin to a bus is a non-sense, because a bus is a collection of signals, and this connection will be ignored by Eeschema.

In the example above, connections are made by the labels placed on wires connected to the pins. Connections via bus entries (wire segments at 45 degrees) to bus wires have only an esthetic value, and are not necessary on the purely schematic level.

In fact, due to the repetition command (*Insert* key), connections can be very quickly made in the following way, if component pins are aligned in increasing order (a common case in practice on components such as memories, microprocessors…):

- Place the first label (for example PCA0)
- Use the repetition command as much as needed to place members. EESchema will automatically create the next labels (PCA1, PCA2…) vertically aligned, theoretically on the position of the other pins.
- Draw the wire under the first label. Then use the repetition command to place the other wires under the labels.
- If needed, place the bus entries by the same way (Place the first entry, then use the repetition command).

*Note:*

In the Preferences/Options menu, you can set the parameters of repetition:

- Vertical step.
- Horizontal step.
- Label increment (which can thus be incremented by 2, 3. or decremented).

### 5.5.3.3 - Global Connections between buses

You may need connections between buses, in order to link two buses having different names, or in the case of a hierarchy, to create connections between different sheets. You can make these connections in the following way.

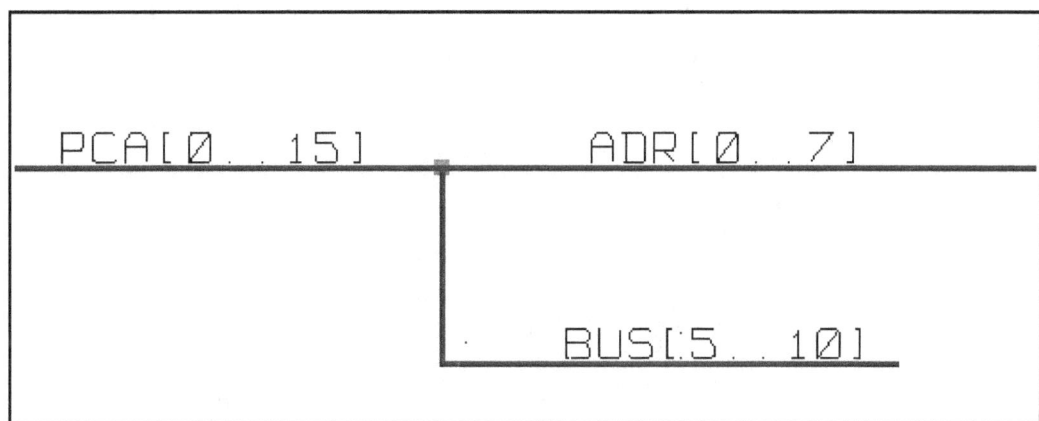

Buses PCA [0..15], ADR [0..7] and BUS [5..10] are connected together (note the junction here because the vertical bus wire joins the middle of the horizontal bus segment).

More precisely, the corresponding members are connected together : PCA0, ADR0 are connected, (as same as PCA1 and ADR1… PCA7 and ADR7).

Furthermore, PCA5, BUS5 and ADR5 are connected (just as PCA6, BUS6 and ADR6 like PCA7, BUS7 and ADR7).

PCA8 and BUS8 are also connected (just as PCA9 and BUS9, PCA10 and BUS10)

On the other hand you cannot connect members of different weights in this way.

If you want to connect members of different weights from different buses, you will have to do that member by member like two usual labels, placing them on the same wire.

### 5.5.4 - Power ports connection

When the power pins of the components are visible, they must be connected, as for any other signal.

The difficulty comes from components (such as gates and flip-flops) for which the power pins are normally invisible (invisible power pins).

The difficulty is double because:

- You cannot connect wires, because of their invisibility.
- You do not know their name.

And moreover, it would be a bad idea to make them visible and to connect them like the other pins, because the schematic would become unreadable and not in accordance with usual conventions.

Note:

If you want to enforce the display of these invisible power pins, you must check the option "Show invisible power pins" in the Preferences/Options dialog box of the main menu, or the icon 🔯 of the left toolbar ( options toolbar )

Eeschema connects automatically the invisible power pins :

All the invisible power pins of the same name are automatically connected between them without other notice.

However these automatic connections must be supplemented:

- By connections to the other visible pins, connected to this power port.

- Possibly by connections between groups of invisible pins of different names (for example, the ground pins are usually called "GND" in TTL components and "VSS" in MOS, and they must be connected together).

For these connections, you must use power ports symbols (components especially designed for this use, that you can create and modify with the library editor).

These symbols consist of an invisible power pin associated with the desired drawing.

Don't use labels, which have only a "local" connection ability, and which would not connect the invisible power pins. (See hierarchy concepts for more details).

The figure below shows an example of power ports connections.

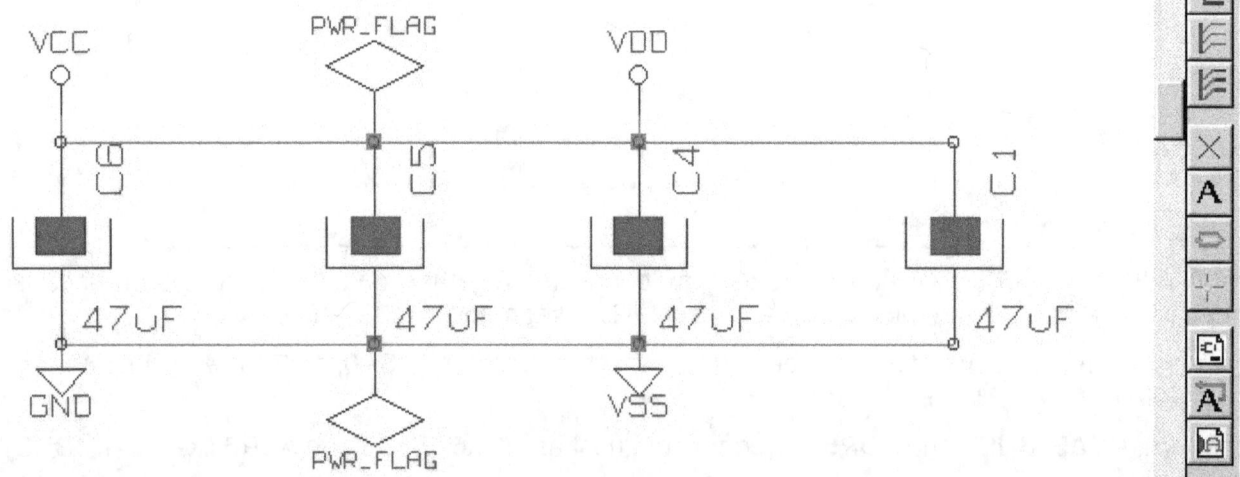

In this example, ground (GND) is connected to power port VSS, and power port VCC is connected to VDD.

Two PWR_FLAG symbols are visible. They indicate that the two power ports VCC and GND are really connected to a power source.

Without these two flags, the ERC tool would diagnose : *Warning: power port not powered*.

All these symbols are components of the schematic library "power".

### 5.5.5 - "No Connection" symbols

These symbols are very useful to avoid undesired warnings in the ERC. The electric rules check ensures that no connection has been inopportunely left unconnected.

If pins must really remain unconnected, it is necessary to place a No-Connection symbol (tool ⊠) on these pins. These symbols however do not have any influence on the generated netlists.

## 5.6 - Drawing Complements

### 5.6.1 - Text Comments

It can be useful (for a good comprehension of the schematic) to place indications such as text fields, frames. Text fields (tool **T**) and dotted lines (tool ◢) are intended for this use, contrary to labels and wires, which are connection elements.

Here you can find an example of a frame with a textual comment.

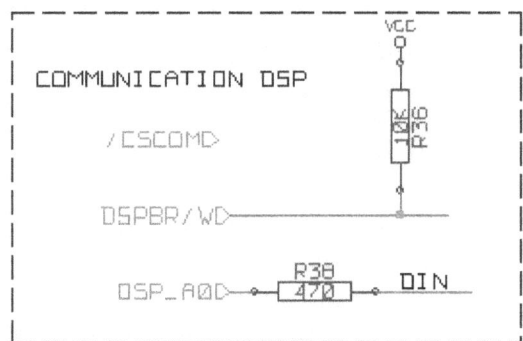

### 5.6.2 - Sheet title block

The title block is edited with the tool 🔲 .

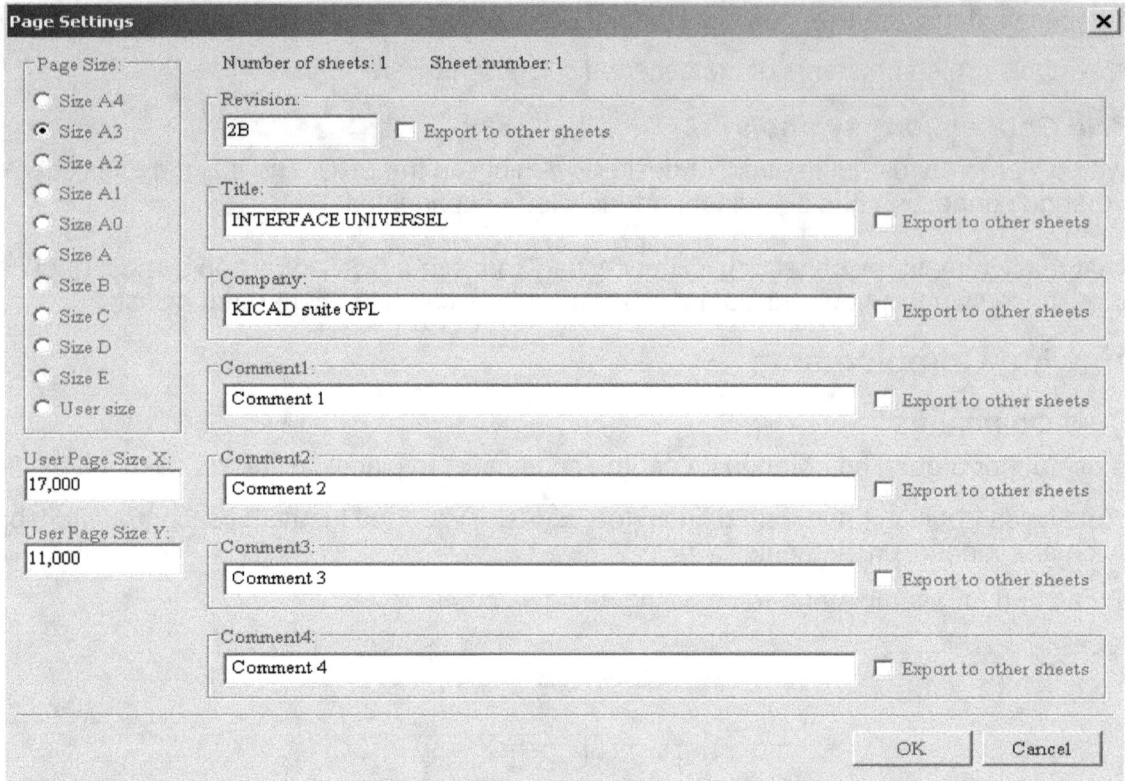

The complete title block will be as follows.

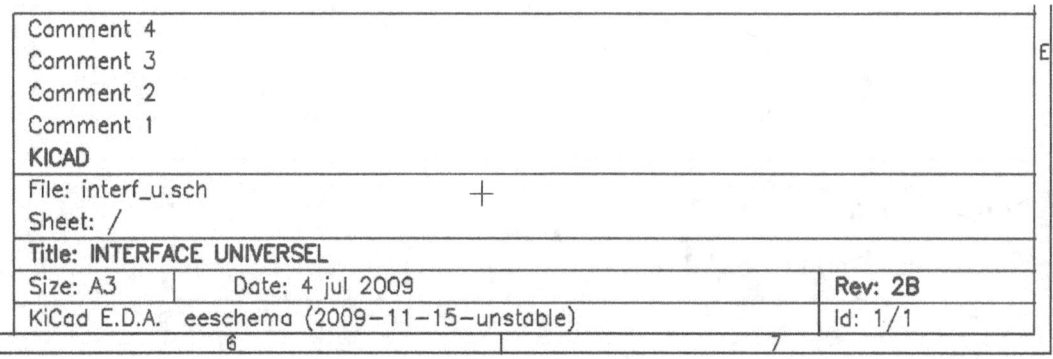

The date and the sheet number (Sheet X/Y) are automatically updated:

- Date : when you modify the schematic.
- Sheet number (useful in hierarchy) : by the annotation function.

# 6 - Hierarchical schematics

## Table of Contents

## 6.1 - Introduction

A hierarchical representation is generally a good solution for projects bigger than a few sheets. If you want to manage this kind of project, it will be necessary to:

- Use large sheets, which results in printing and handling problems.

- Use several sheets, which leads you to a hierarchy structure.

The complete schematic then consists in a main schematic sheet, called root sheet, and sub-sheets constituting the hierarchy. Moreover, a skillful subdividing of the design into separate sheets often improves on its readability.

From the root sheet, you must be able to find all sub-sheets. Hierarchical schematics management is

very easy with Eeschema, thanks to an integrated "hierarchy navigator" accessible via the icon of the upper and right toolbar.

There are two types of hierarchy that can exist simultaneously: the first one has just been evoked and is of general use. The second consists in creating components in the library that appear like traditional components in the schematic, but which actually correspond to a schematic which describes their internal structure.

This second type is used to develop integrated circuits, because in this case you have to use function libraries in the schematic you are drawing.

Eeschema currently doesn't treat this second case.

A hierarchy can be:

- simple: a given sheet is used only once

- complex: a given sheet is used more than once (multiples instances)

- Flat: which is a simple hierarchy, but connections between sheets are not drawn.

Eeschema can deal with all these hierarchies.

The creation of a hierarchical schematic is easy, the whole hierarchy is handled starting from the root schematic, as if you had only one schematic.

The two important steps to understand are:

- How to create a sub-sheet.
- How to build electric connections between sub-sheets.

## 6.2 - Navigation in the Hierarchy

Navigation among sub-sheets It is very easy thanks to the navigator tool accessible via the button

 on the horizontal toolbar.

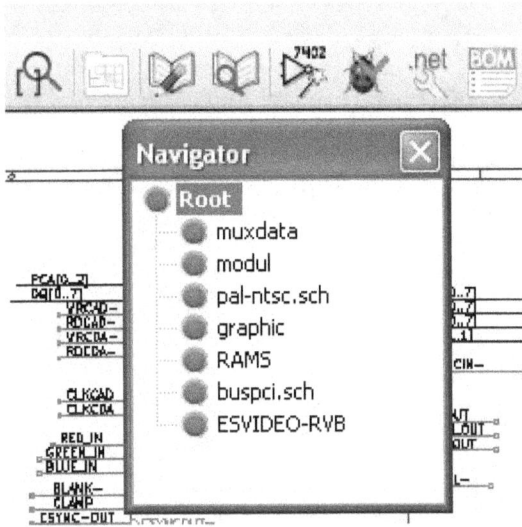

Each sheet is reachable by clicking on its name. For quick access, right click on a sheet name, and choose to enter into sheet.

You can quickly reach the root sheet, or a sub-sheet thanks to the tool ⊞ of the right vertical toolbar. After the navigation tool has been selected:

- Click on a sheet name to selection this sheet.
- Click elsewhere to select the main sheet.

## 6.3 - Local, hierarchical and global labels

### 6.3.1 - Properties

Local labels, tool **A**, are connecting signals only within a sheet. Hierarchical labels (tool **A**) are connecting signals only within a sheet and to a hierarchical pin placed in the parent sheet.

Global labels (tool **A**) are connecting signals across all the hierarchy. Power pins (type *power in* and *power out*) invisible are like global labels because they are seen as connected between them across all the hierarchy.

### 6.3.2 - Notes

Within a hierarchy (simple or complex) one can use both hierarchical labels and/or global labels.

## 6.4 - Hierarchy creation of headlines

You have to:

- Place in the root sheet a hierarchy symbol called "sheet symbol".
- Enter into the new schematic (sub-sheet) with the navigator and draw it, like any other schematic.
- Draw the electric connections between the two schematics by placing Global Labels (HLabels) in the new schematic (sub-sheet), and labels having the same name in the root sheet, known as

SheetLabels. These SheetLabels will be connected to the sheet symbol of the root sheet to the other elements of the schematic like standard component pins.

## 6.5 - Sheet symbol

Draw a rectangle defined by two diagonal points symbolizing the sub-sheet.

The size of this rectangle must allow you to place later particular labels, hierarchy pins, corresponding to the global labels (HLabels) in the sub-sheet .

These labels are similar to usual component pins. Select the tool 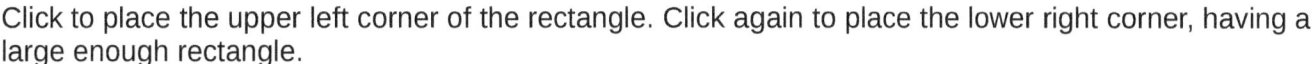 .

Click to place the upper left corner of the rectangle. Click again to place the lower right corner, having a large enough rectangle.

Example :

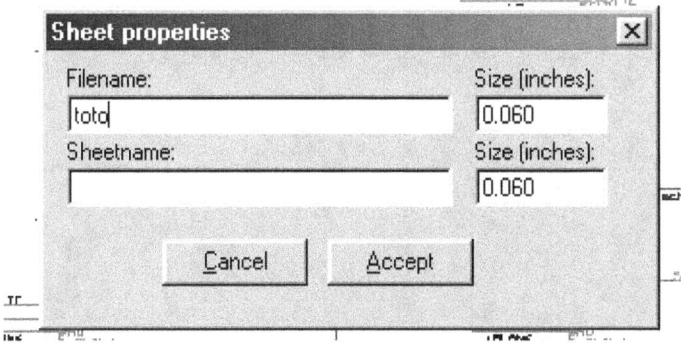

You will then be prompted to type a file name and a sheet name for this sub-sheet (in order to reach the corresponding schematic, using the hierarchy navigator).

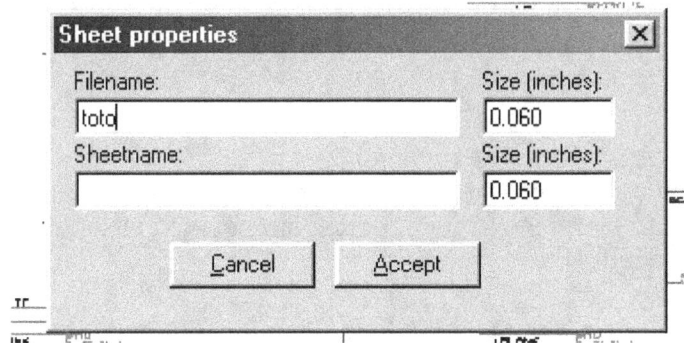

You must give at least a file name. If there is no sheet name, the file name will be used as sheet name (usual way to do that).

## 6.6 - Connections – hierarchical pins

You will create here points of connection (hierarchy pins) for the symbol which has been just created.

These points of connection are similar to normal component pins, with however the possibility to connect a complete bus with only one point of connection.

There are two ways to do this:

- Place the different pins before drawing the sub-sheet (manual placement).
- Place the different pins after drawing the sub-sheet, and the global labels (semi-automatic placement).

The second solution is quite preferable.

**Manual placement:**

- To select the tool  .

- Click on the hierarchy symbol where you want to place this pin.

See below an example of the creation of the hierarchical pin called "CONNEXION".

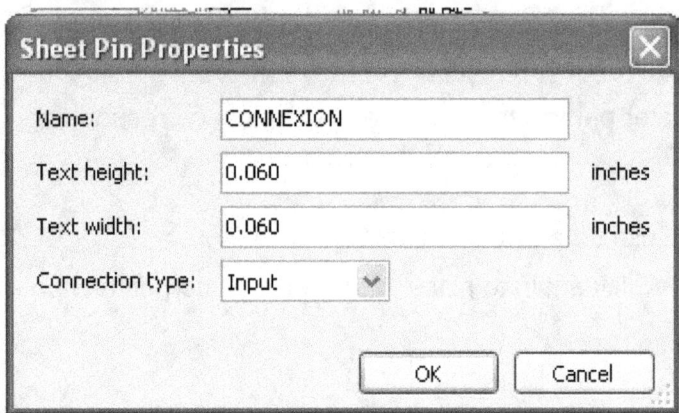

You can define its graphical attributes, and size or later, by editing this pin sheet ( Right click and select Edit in the PopUp menu).

Various pin symbols are available :

- Input
- Output
- BiDir
- Tri State
- Not Specified

These pin symbols are only graphic enhancements, and have no other role.

**Automatic placement:**

- Select the tool  .
- Click on the hierarchy symbol from where you want to import the pins corresponding to global labels placed in the corresponding schematic. A hierarchical pin appears, if a new global label exists, i.e. not corresponding to an already placed pin.
- Click where you want to place this pin.

All necessary pins can thus be placed quickly and without error. Their aspect is in accordance with corresponding global labels.

## 6.7 - Connections - hierarchical labels

Each pin of the sheet symbol just created, must correspond to a label called hierarchical Label in the sub-sheet. Hierarchical labels are similar to labels, but they provide connections between sub-sheet and root sheet. The graphical representation of the two complementary labels (pin and HLabel) is

similar. Hierarchical labels creation is made with the tool ⬛ .

See below a root sheet example:

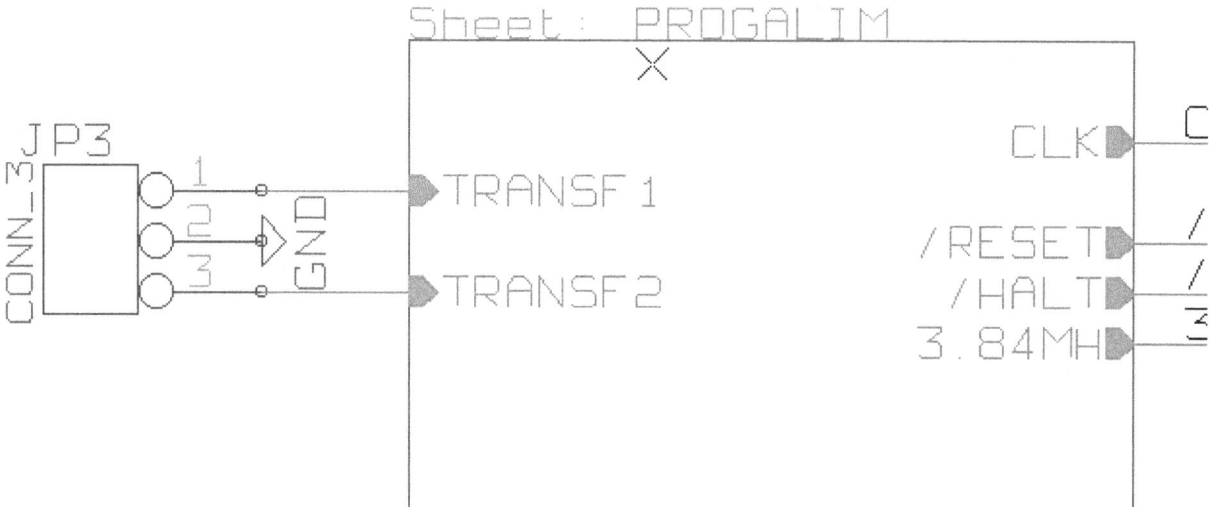

Notice pins TRANSF1 and TRANSF2, connected to connector JP3.

Here are the corresponding connections in the sub-sheet :

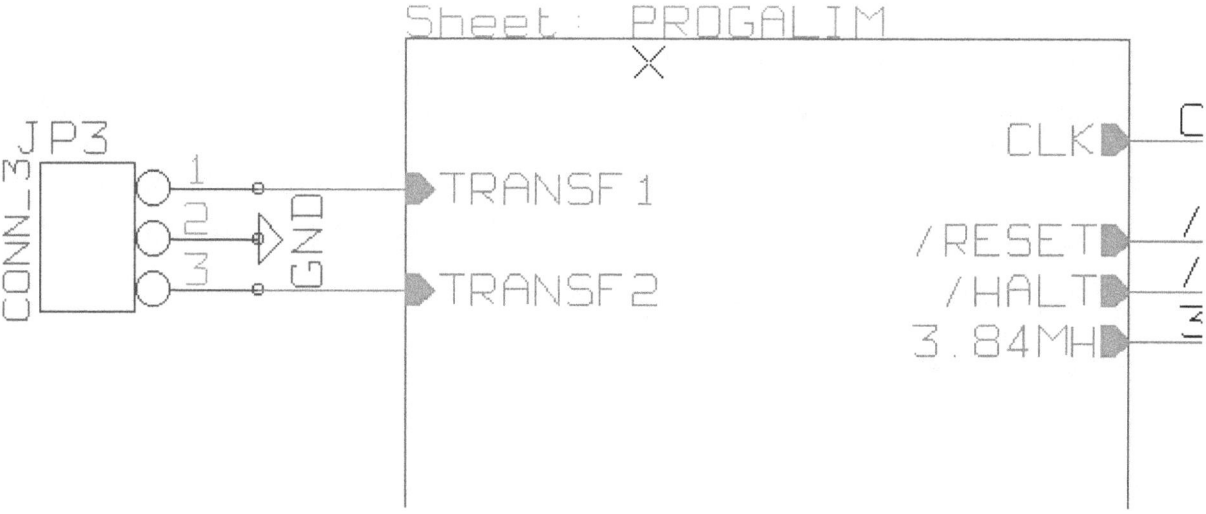

You find again, the two corresponding hierarchical labels, providing connection between the two hierarchical sheets.

**Note**

You can use hierarchical labels and hierarchy pins to connect two buses, according to the syntax (Bus [N. .m]) previously described.

### 6.7.1 - Labels, hierarchical labels, global labels and invisible power pins

Here are some comments on various ways to provide connections, others than wire connections.

#### 6.7.1.1 - Simple labels

Simple labels have a local capacity of connection, i.e. limited to the schematic sheet where they are placed. This is due to the fact that :

- Each sheet has a sheet number.
- This sheet number is associated to a label.

Thus, if you place the label "TOTO" in sheet n° 3, in fact the true label is "TOTO_3". If you also place a label "TOTO" in sheet n° 1 ( root sheet) you place in fact a label called "TOTO_1", different from "TOTO_3". This is always true, even if there is only one sheet.

### 6.7.1.2 - Hierarchical labels

What is said for the simple labels is also true for hierarchical labels.

Thus in the same sheet, a HLabel "TOTO" is considered to be connected to a local label "TOTO", but not connected to a HLabel or label called "TOTO" in another sheet.

However a HLabel is considered to be connected to the corresponding SheetLabel symbol in the hierarchical symbol placed in the root sheet.

### 6.7.1.3 - Invisible power pins

It was seen that invisible power pins were connected together if they have the same name. Thus all the power pins declared "Invisible Power Pins" and named VCC are connected and form the equipotential VCC, whatever the sheet they are placed on.

This means that if you place a VCC label in a sub-sheet, it will not be connected to VCC pins, because this label is actually VCC_n, where n is the sheet number.

If you want this label VCC to be really connected to the equipotential VCC, it will have to be explicitly connected to an invisible power pin, thanks to a VCC power port.

### 6.7.2 - Global labels

Global labels that have an identical name are connected across the whole hierarchy.

(power labels like vcc ... are global labels)

## 6.8 - Complex Hierarchy

Here is an example. The same schematic is used twice (two instances). The two sheets share the same schematic because the file name is the same for the two sheets ("other_sheet.sch"). But the sheet names must be different.

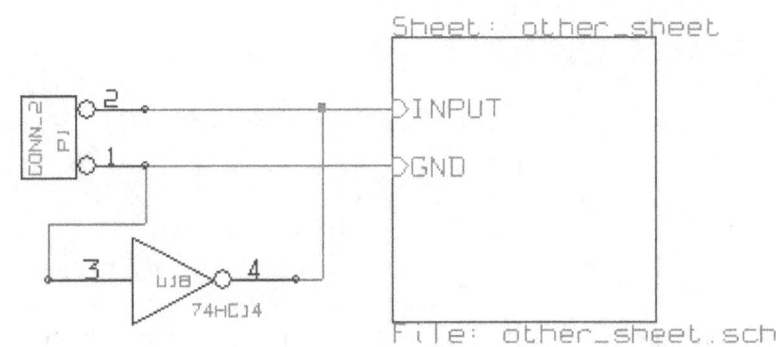

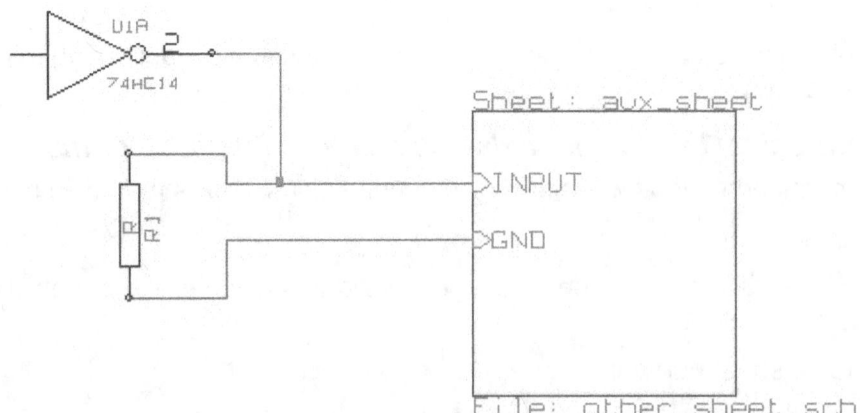

## 6.9 - Flat hierarchy

You can create a project using many sheets, without creating connections between these sheets (flat hierarchy) if the next rules are repsected:

- You must create a root sheet containing the other sheets, which acts as a link between others sheets.

- No explicit connections are needed.

- All connections between sheets will use global labels instead of hierarchical labels.

Here is an example of a root sheet.

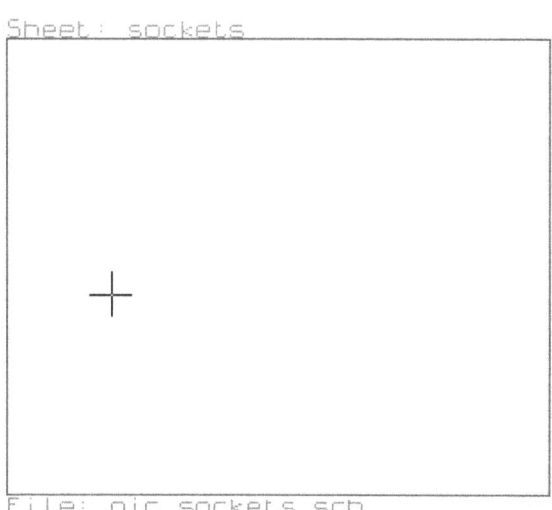

Here is the two pages, connected by global labels.

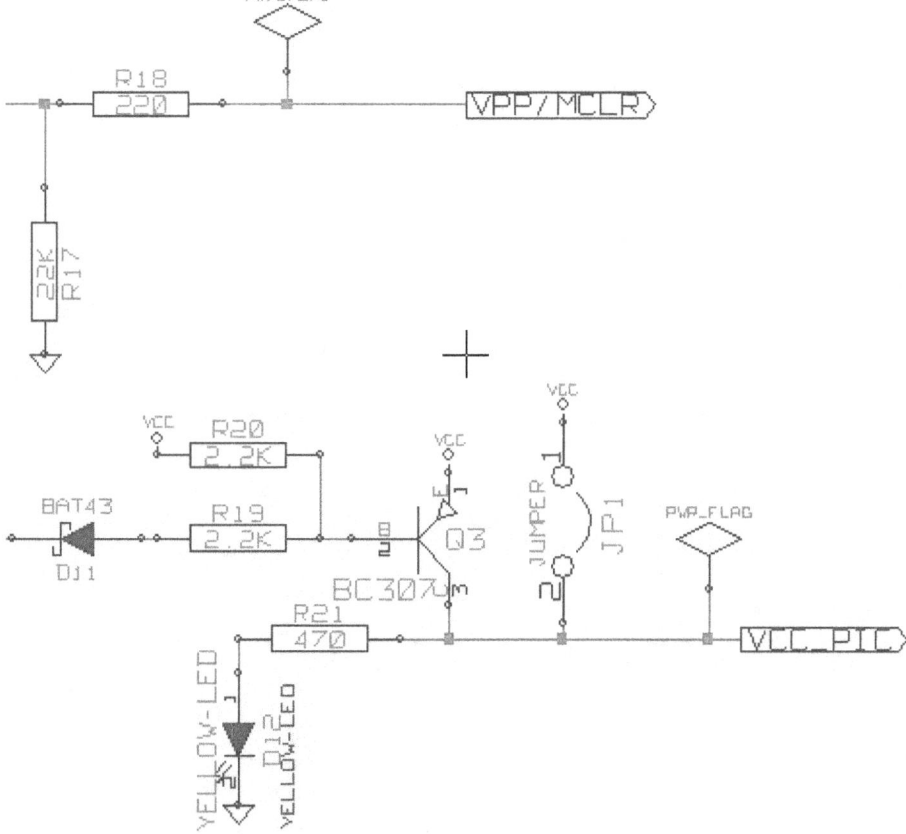

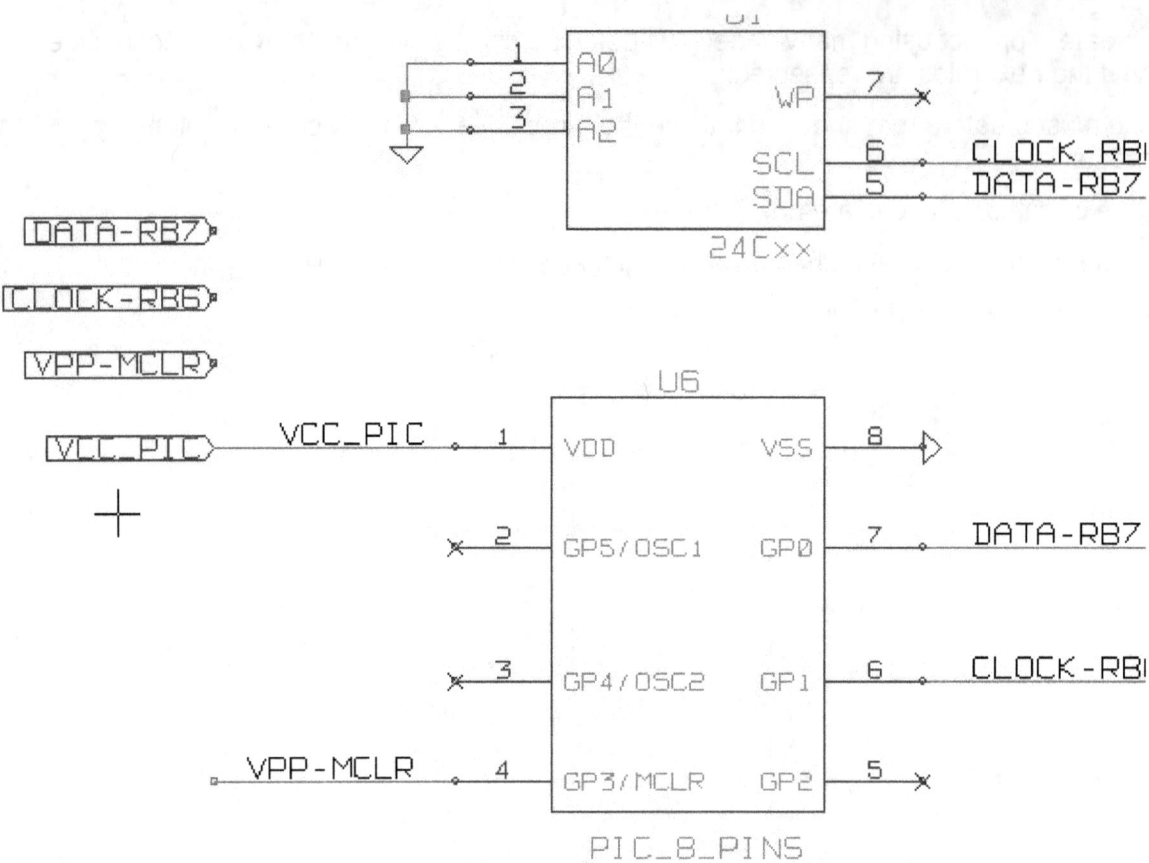

| | DATA-RB7 |
|---|---|
| Look at global labels | CLOCK-RB6 |
| | VPP-MCLR |

# 7 - Automatic classification Annotation

## Table of Contents

## 7.1 - Introduction

The automatic classification annotation tool allows you to automatically assign a designator to components in your schematic. For multi-parts components, assign a multi-part suffix to minimize the

number of these packages. The automatic classification annotation tool is accessible via the icon . Here you find its main window.

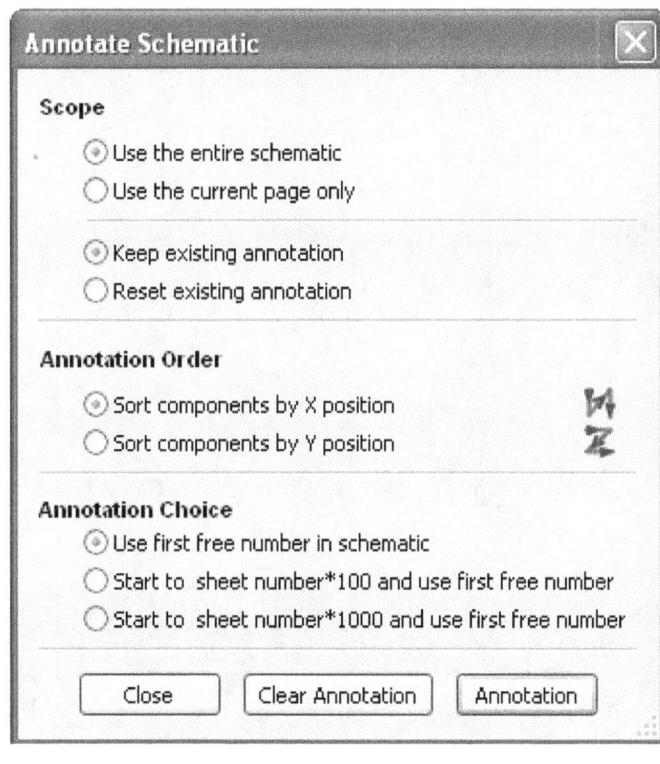

Various possibilities are available:

- Annotate all the components (reset existing annotation option)
- Annotate new components only (i.e. those whose reference finishes by? like IC? ) (keep existing annotation option).
- Annotate the whole hierarchy (use the entire schematic option).
- Annotate the current sheet only (use current page only option).

The annotation order choice gives the method used to set the reference number inside each sheet of the hierarchy.

Except for particular cases, an automatic annotation applies to the whole project (all sheets) and to the new components, if you don't want to modify previous annotations.

The Annotation Choice gives the method used to calculate reference Id:

- Use first free number in schematic: components are annotated from 1 (for each reference prefix). If a previous annotation exists, not yet in use numbers will be used.

- Start to sheet number*100 and use first free number:
  Annotation start from 101 for the sheet 1, from 201 for the sheet 2, etc.
  If there are more than 99 items having the same reference prefix (U, R) inside the sheet 1, the annotation tool uses the number 200 and more, and annotation for sheet 2 will start from the next free number.

- Start to sheet number*1000 and use first free number.
  Annotation start from 1001 for the sheet 1, from 2001 for the sheet 2.

## 7.2 - Some examples

### 7.2.1 - Annotation order

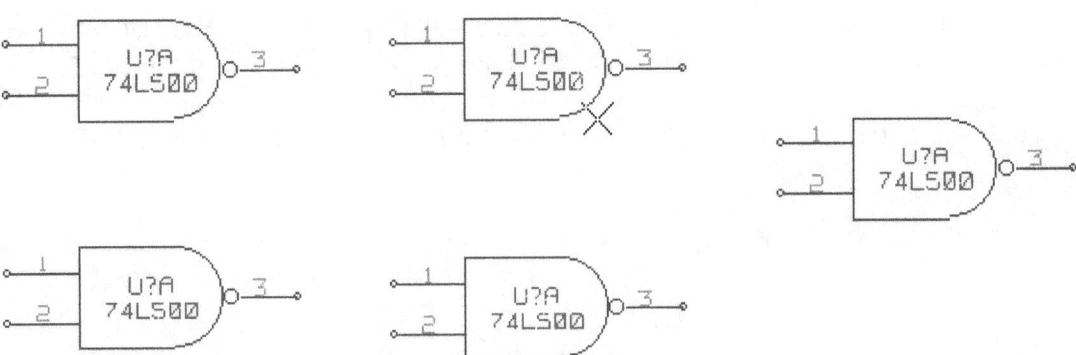

This example shows 5 elements placed, but not annotated.

After the annotation tool Is executed, the following result is obtained.

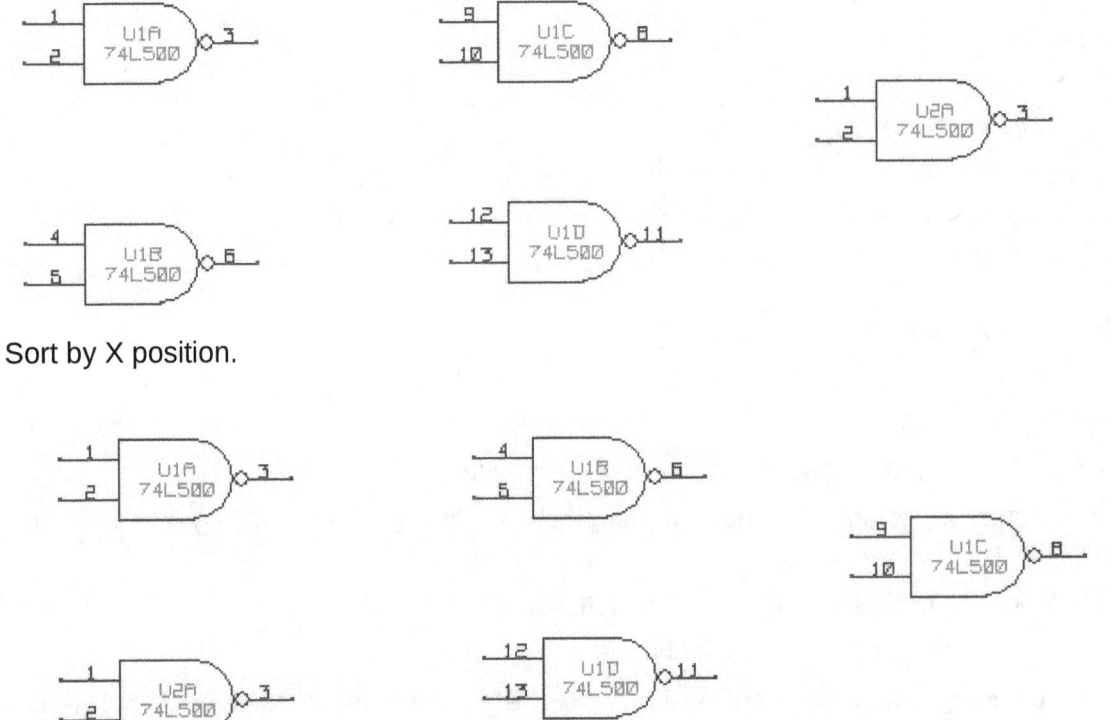

Sort by X position.

Sort by Y position.

You can see that four 74LS00 gates were distributed in U1 package, and that the fifth 74LS00 has been assigned to the next , U2.

## 7.2.2 - Annotation Choice

Here is an annotation in sheet 2 where the option use first free number in schematic was set.

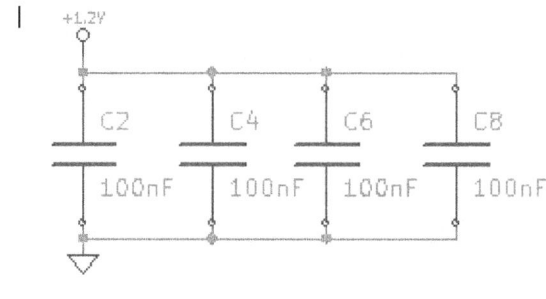

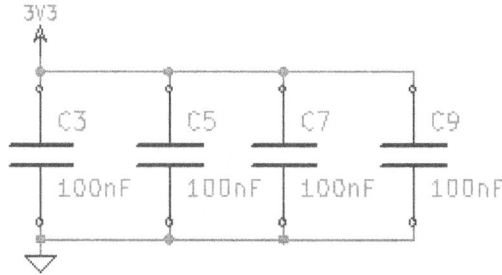

Option start to sheet number*100 and use first free number give the following result.

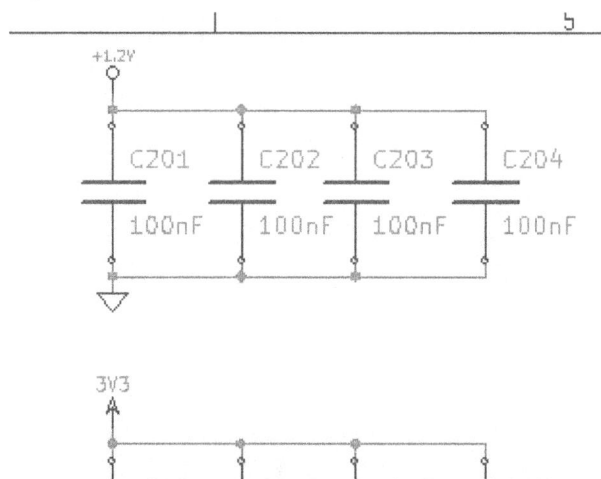

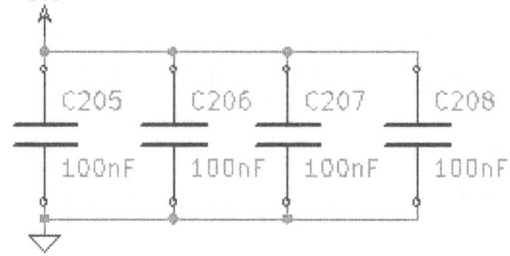

The option start to sheet number*1000 and use first free number gives the following result.

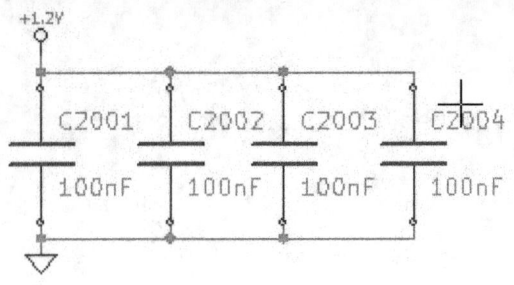

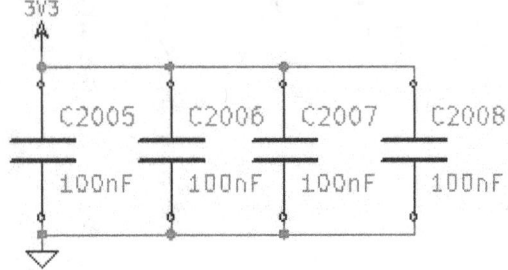

# 8 - Design verification with electrical rules check

## Table of Contents

## 8.1 - Introduction

The Electrical Rules Check (ERC) tool performs an automatic check of your schematic. The ERC checks for any errors in your sheet, such as unconnected pins, unconnected hierarchical symbols, shorted outputs, etc. Naturally, an automatic check is not infallible, and the software that make it possible to detect all design errors is not yet 100% complete. Such a check is very useful, because it allows you to detect many oversights and small errors.

In fact all detected errors must be checked and then corrected before proceeding as normal. The quality of the ERC is directly related to the care taken in declaring electrical pin properties during library creation. ERC output is reported as "errors" or "warnings".

```
┌─────────────────────────────────────────────────────────────────────────┐
│ EESchema Erc                                                         [X]  │
│ ┌──────┬─────────┐                                                        │
│ │ ERC  │ Options │                                                        │
│ ┌┴──────┴──────────────────┐  Messages:              ┌──────────────┐    │
│ │ Erc File Report:          │  ┌──────────────────┐   │   Test Erc   │    │
│ │                           │  │                 ∧│   └──────────────┘    │
│ │  Total Errors Count:   2  │  │                  │   ┌──────────────┐    │
│ │                           │  │                  │   │  Del Markers │    │
│ │  Warnings Count:       1  │  │                  │   └──────────────┘    │
│ │                           │  │                 ∨│   ┌──────────────┐    │
│ │  Errors Count:         1  │  └──────────────────┘   │    Close     │    │
│ └───────────────────────────┘                         └──────────────┘    │
│   ☐ Create ERC report                                                     │
│                                                                           │
│   Markers:                                                                │
│  ┌──────────────────────────────────────────────────────────────────────┐│
│  │ ErrType(5): Conflict problem between pins. Severity: error            ││
│  │    ◆ @ (7,7000 ",4,9500 "): Cmp U10, Pin 3 (output) connected to      ││
│  │    ◆ @ (7,7000 ",5,5500 "): Cmp U10, Pin 6 (output) (net 179)         ││
│  │ ErrType(4): Conflict problem between pins. Severity: warning          ││
│  │    ◆ @ (10,1000 ",3,3000 "): Cmp U9, Pin B12 (BiDi) connected to      ││
│  │    ◆ @ (3,2500 ",8,9000 "): Cmp #PWR09, Pin 1 (power_out) (net 10)     ││
│  │                                                                        ││
│  │                                                                        ││
│  │                                                                        ││
│  │                                                                        ││
│  │                                                                        ││
│  │                                                                        ││
│  │                                                                        ││
│  │                                                                        ││
│  └──────────────────────────────────────────────────────────────────────┘│
└─────────────────────────────────────────────────────────────────────────┘
```

## 8.2 - How to use ERC

ERC can be started by clicking on the icon  .

Warnings are placed on the schematic elements rising an ERC error (pins, or labels).

Notes:

- In this dialog window, when clicking on an error message you can jump to the corresponding marker in schematic.

- In the schematic right click on a marker to access the corresponding diagnostic message.

You can also delete error markers from the dialog.

## 8.3 - Example of ERC

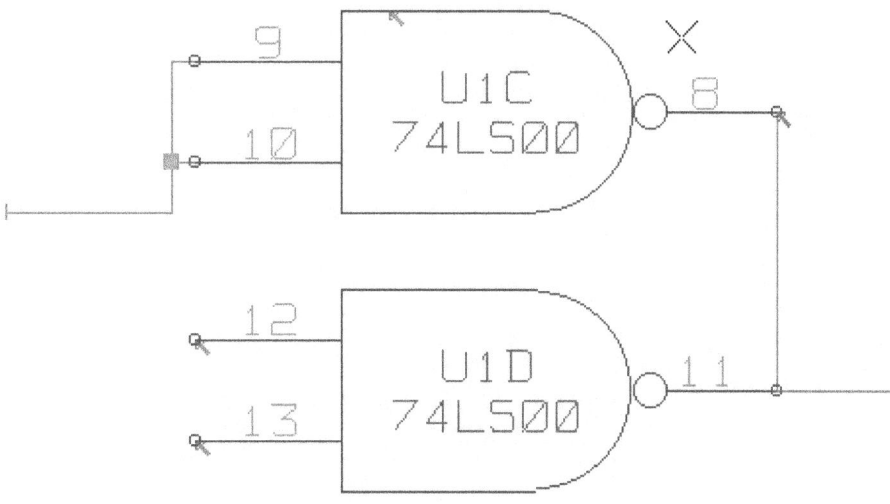

Here you can see four errors:

- Two outputs have been erroneously connected together (red arrow).
- Two inputs have been left unconnected (green arrow).
- There is an error on an invisible power port, power flag is missing (green arrow on the top).

## 8.4 - Displaying diagnostics

By right clicking on a marker the pop menu allows to access the ERC marker diagnostic window.

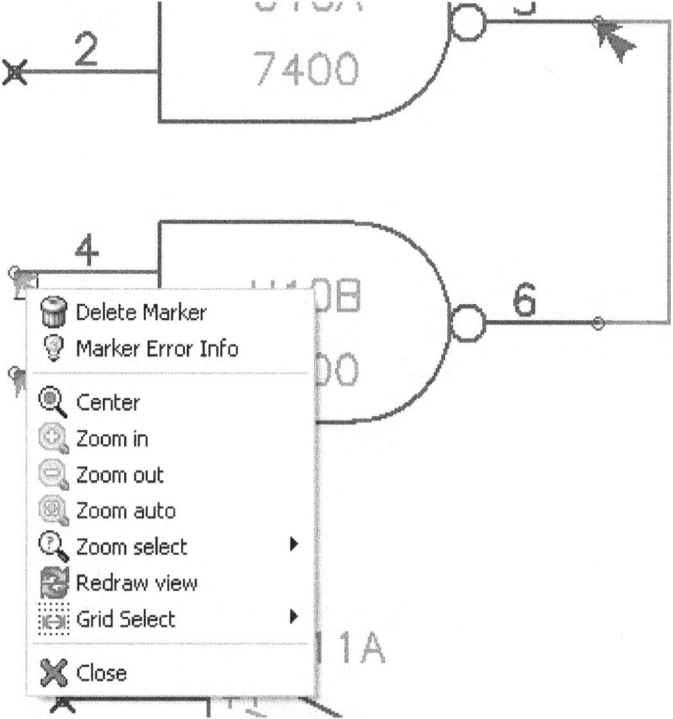

and when clicking on Marker Error Info you can get a description of the error.

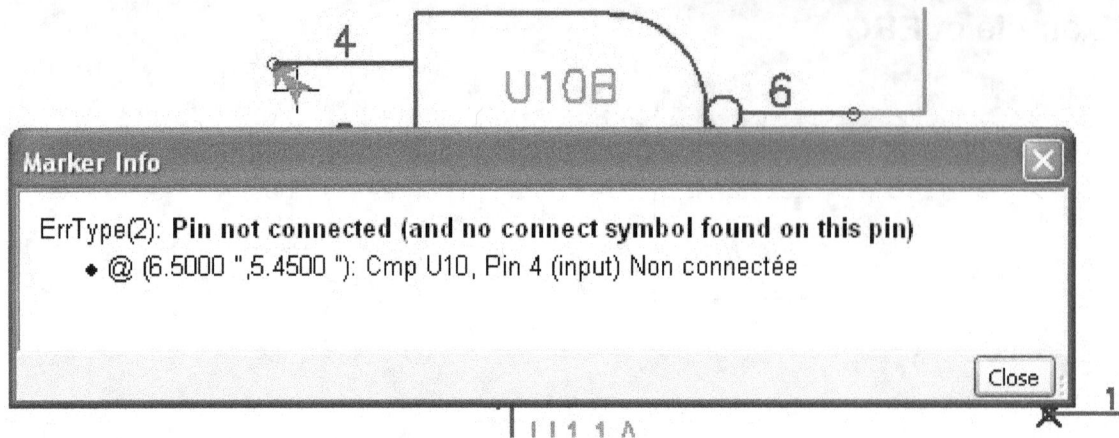

## 8.5 - Power flags and Power flags

It is common to have an error or a warning on power pins, even though all seems normal. See example above. This happens because, in most designs, the power is provided by connectors, that are not power sources (like regulator output, which is declared as Power out).

The ERC thus won't detect any Power out pin to control this wire and will declare them not driven by a power source.

To avoid this warning you have to place a "PWR_FLAG" on such a power port. Take a look at the following example.

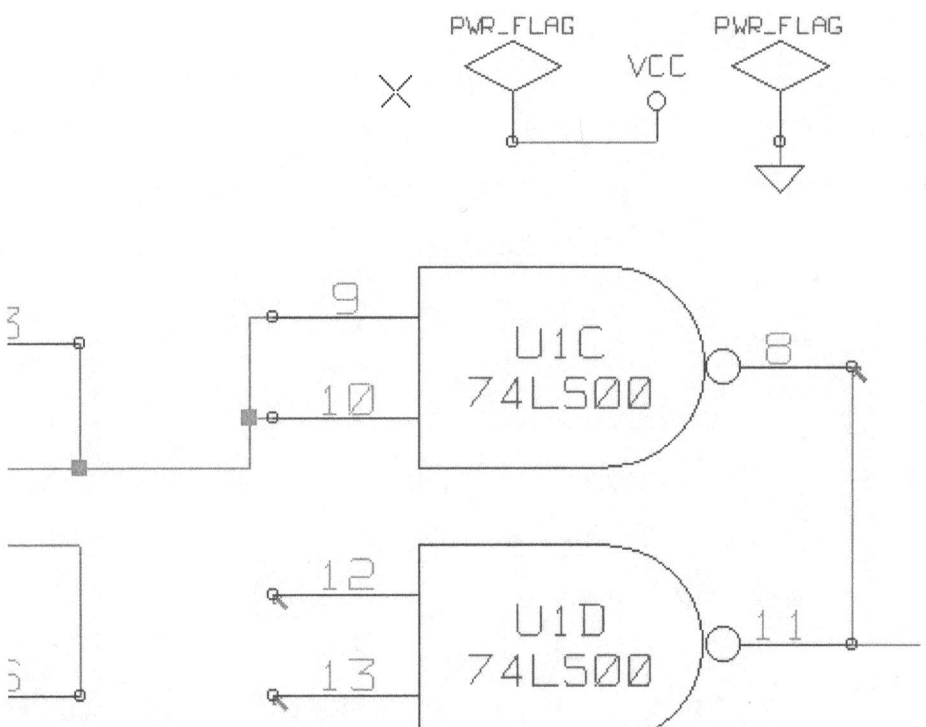

The error marker will then disappear.

Most of the time, a PWR_FLAG must be connected to GND, because usually regulators have outputs declared as power out, but ground pins are never power out (the normal attribute is power in), so grounds never appear connected to a power source without a pwr_flag.

## 8.6 - Configuration

The Options panel allows you to configure connectivity rules to define electrical conditions for errors and warnings check.

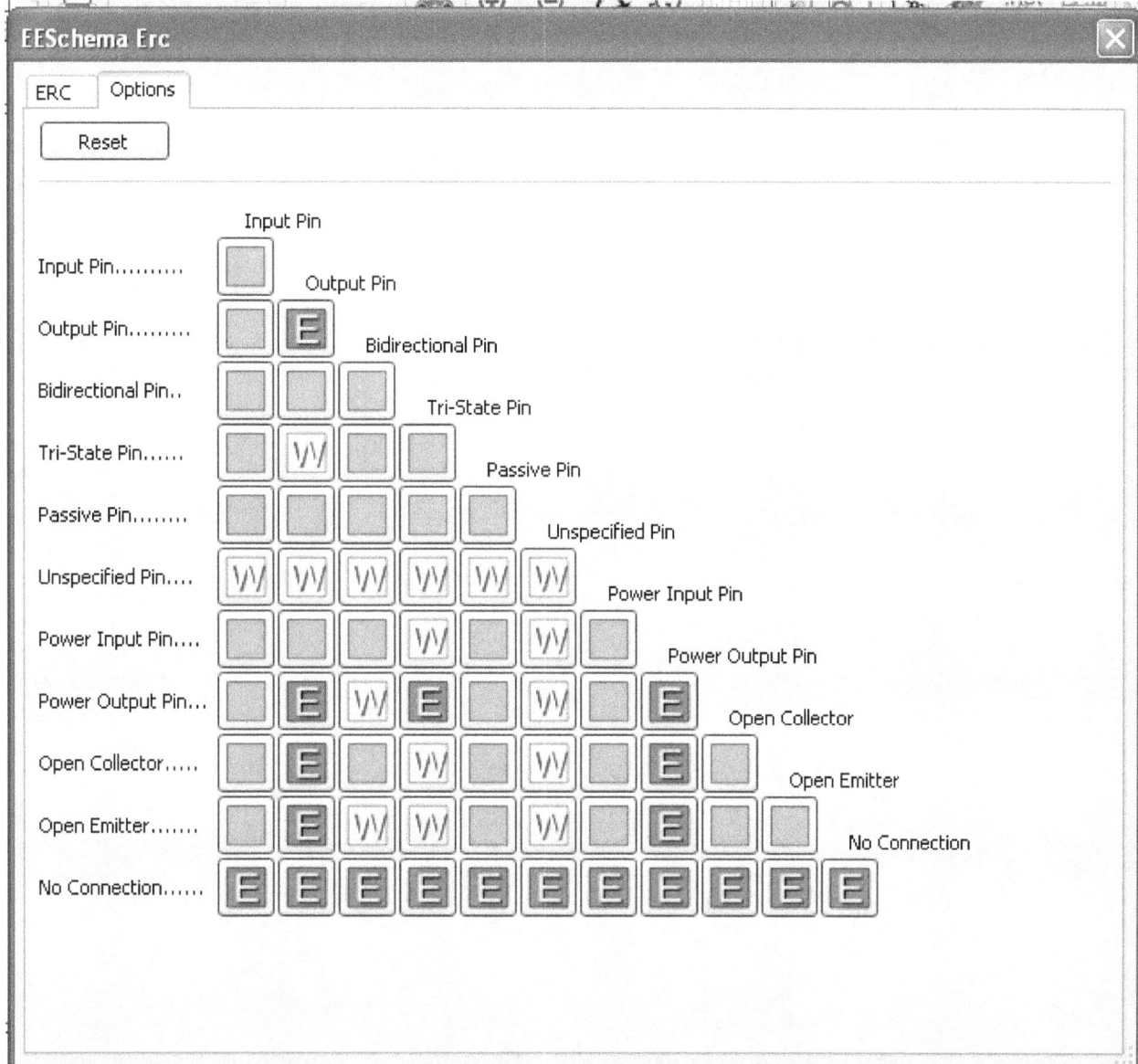

Rules can be changed by clicking on the desired square of the matrix, causing it to cycle through the choices : normal, warning, error.

## 8.7 - ERC report file

An ERC report file can be generated and saved by checking the option Write ERC report. The file extension for ERC report files is .erc. Here is an example of ERC report file.

```
ERC control (4/1/1997-14:16:4)

***** Sheet 1 (INTERFACE UNIVERSAL)
ERC: Warning Pin input Unconnected @ 8.450, 2.350
ERC: Warning passive Pin Unconnected @ 8.450, 1.950
ERC: Warning: BiDir Pin connected to power Pin (Net 6) @ 10.100, 3.300
ERC: Warning: Power Pin connected to BiDir Pin (Net 6) @ 4.950, 1.400

>> Errors ERC: 4
```

# 9 - Create a Netlist

## Table of Contents

## 9.1 - Overview

A netlist is a file which describes electrical connections between components. In the netlist file you can find:

- The list of the components
- The list of connections between components, called equip-potential nets.

Different netlist formats exist. Sometimes the components list and the equi-potential list are two separate files. This netlist is fundamental in the use of schematic capture software, because the netlist is the link with other electronic CAD software, like:

- PCB software.
- Schematic and PCB Simulators.
- CPLD (and other programmable IC's) compilers.

Eeschema supports several netlist formats.

- PCBNEW format (printed circuits).
- ORCAD PCB2 format (printed circuits).
- CADSTAR format (printed circuits).
- Spice format, for various simulators (the Spice format is also used by other simulators).

## 9.2 - Netlist format

Select the tool ![.net] to open the netlist creation dialog box.

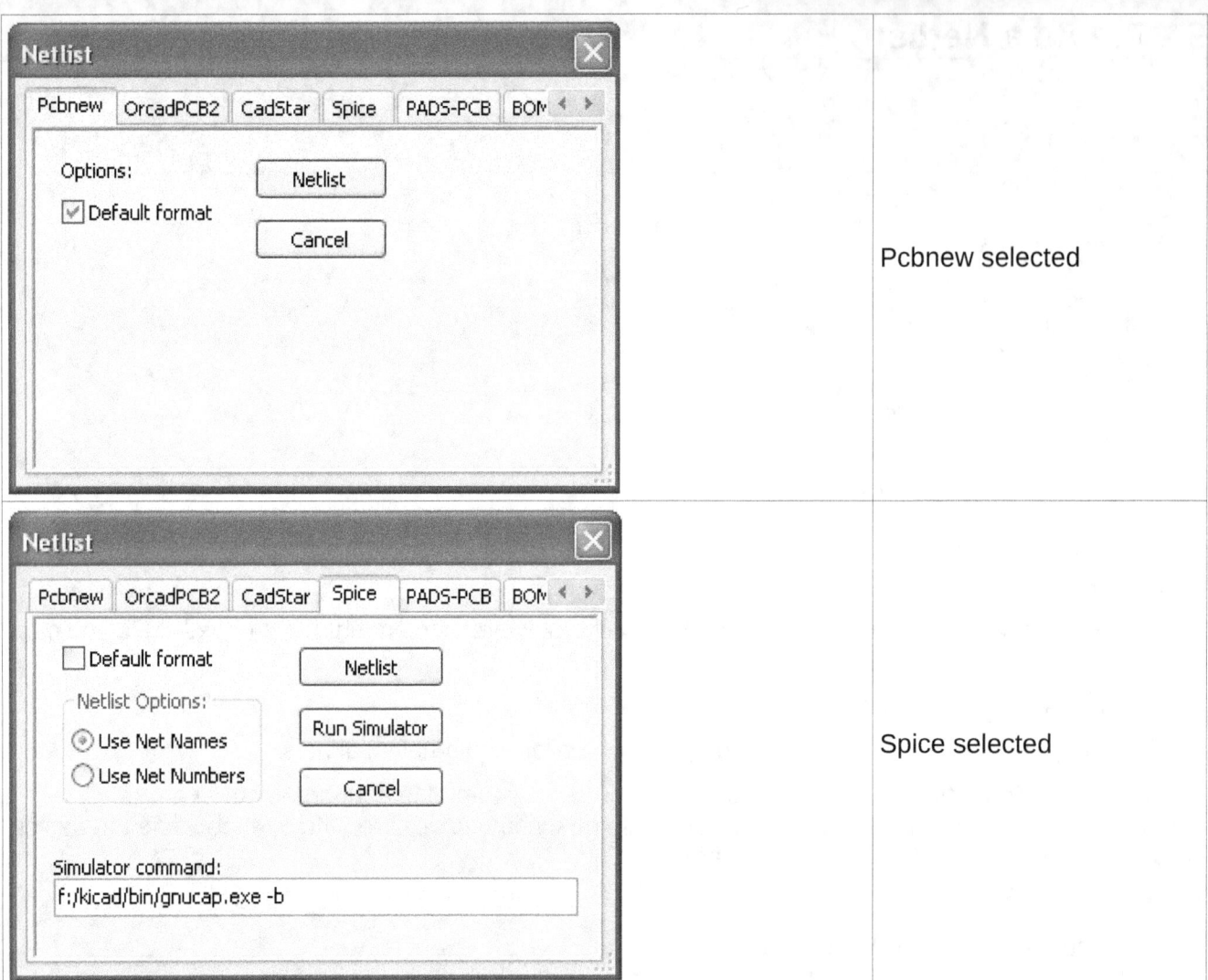

| | |
|---|---|
| (Netlist dialog — Pcbnew tab: Options: Default format checked; Netlist, Cancel buttons) | Pcbnew selected |
| (Netlist dialog — Spice tab: Default format unchecked; Netlist Options: Use Net Names, Use Net Numbers; Netlist, Run Simulator, Cancel buttons; Simulator command: f:/kicad/bin/gnucap.exe -b) | Spice selected |

Using the different tabs you can select the desired format. In Spice format you can generate netlists with either equi-potential names (it is more legible) or net numbers (old Spice versions accept numbers only). By clicking the Netlist button, you will be asked for a netlist file name.

*Note*

With big projects, the netlist generation can take up to few minutes.

## 9.3 - Netlist Examples

You can see below a schematic design using the PSPICE library.

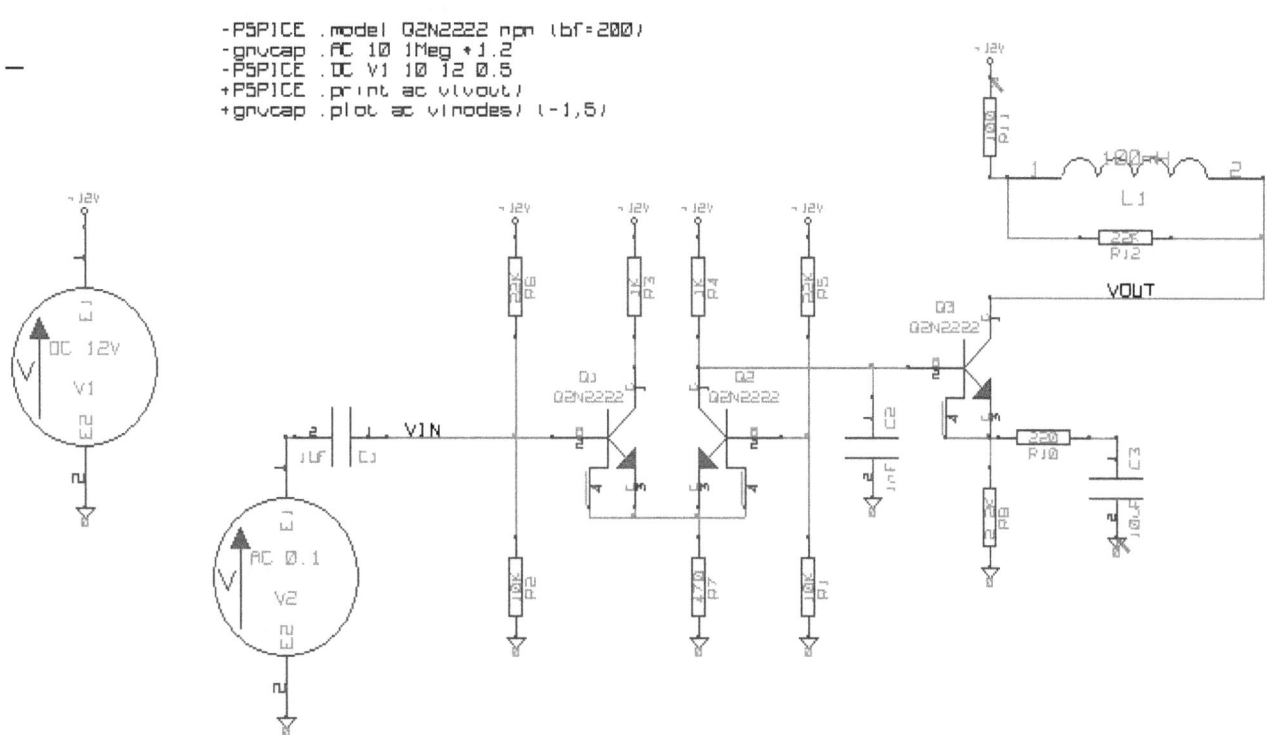

Example of a PCBNEW netlist file.

```
# EESchema Netlist Version 1.0 generee le 21/1/1997-16:51:15
(
(32E35B76 $noname C2 1NF {Lib=C}
(1 0)
(2 VOUT_1)
)
(32CFC454 $noname V2 AC_0.1 {Lib=VSOURCE}
(1 N-000003)
(2 0)
)
(32CFC413 $noname C1 1UF {Lib=C}
(1 INPUT_1)
(2 N-000003)
)
(32CFC337 $noname V1 DC_12V {Lib=VSOURCE}
(1 +12V)
(2 0)
)
(32CFC293 $noname R2 10K {Lib=R}
(1 INPUT_1)
(2 0)
)
(32CFC288 $noname R6 22K {Lib=R}
(1 +12V)
(2 INPUT_1)
)
(32CFC27F $noname R5 22K {Lib=R}
(1 +12V)
(2 N-000008)
)
(32CFC277 $noname R1 10K {Lib=R}
(1 N-000008)
(2 0)
)
(32CFC25A $noname R7 470 {Lib=R}
(1 EMET_1)
```

```
(2 0)
)
(32CFC254 $noname R4 1K {Lib=R}
(1 +12V)
(2 VOUT_1)
)
(32CFC24C $noname R3 1K {Lib=R}
(1 +12V)
(2 N-000006)
)
(32CFC230 $noname Q2 Q2N2222 {Lib=NPN}
(1 VOUT_1)
(2 N-000008)
(3 EMET_1)
)
(32CFC227 $noname Q1 Q2N2222 {Lib=NPN}
(1 N-000006)
(2 INPUT_1)
(3 EMET_1)
)
)
# End
```

In PSPICE format, the netlist is as follows.

```
* EESchema Netlist Version 1.1 (Spice format) creation date: 18/6/2008-08:38:03

.model Q2N2222 npn (bf=200)
.AC 10 1Meg *1.2
.DC V1 10 12 0.5

R12  /VOUT N-000003 22K
R11  +12V N-000003 100
L1   N-000003 /VOUT 100mH
R10  N-000005 N-000004 220
C3   N-000005 0 10uF
C2   N-000009 0 1nF
R8   N-000004 0 2.2K
Q3   /VOUT N-000009 N-000004 N-000004 Q2N2222
V2   N-000008 0 AC 0.1
C1   /VIN N-000008 1UF
V1   +12V 0 DC 12V
R2   /VIN 0 10K
R6   +12V /VIN 22K
R5   +12V N-000012 22K
R1   N-000012 0 10K
R7   N-000007 0 470
R4   +12V N-000009 1K
R3   +12V N-000010 1K
Q2   N-000009 N-000012 N-000007 N-000007 Q2N2222
Q1   N-000010 /VIN N-000007 N-000007 Q2N2222

.print ac v(vout)
.plot ac v(nodes) (-1,5)

.end
```

## 9.4 - Note

### 9.4.1 - Netlist Name Precautions

Many software tools that use netlists do not accept spaces in the component names, pins, equipotentials or others. Systematically avoid spaces in labels, or names and value fields of components or their pins.

In the same way, special characters other than letters and numbers can induce problems. Note that this limitation is not related to Eeschema, but to the netlist formats that can then become not translatable to softwares that use netlist files.

### 9.4.2 - PSPICE netlists

For the Pspice simulator, you have to include some command lines in the netlist itself (.PROBE, .AC, etc.).

Any text line included in the schematic diagram starting with the keyword **-pspice** or **-gnucap** will be inserted (without the keyword) at the top of the netlist.

Any text line included in the schematic diagram starting with the keyword **+pspice** or **+gnucap** will be inserted (without the keyword) at the end of the netlist.

Here is a sample using many one line texts and one multi-line text.

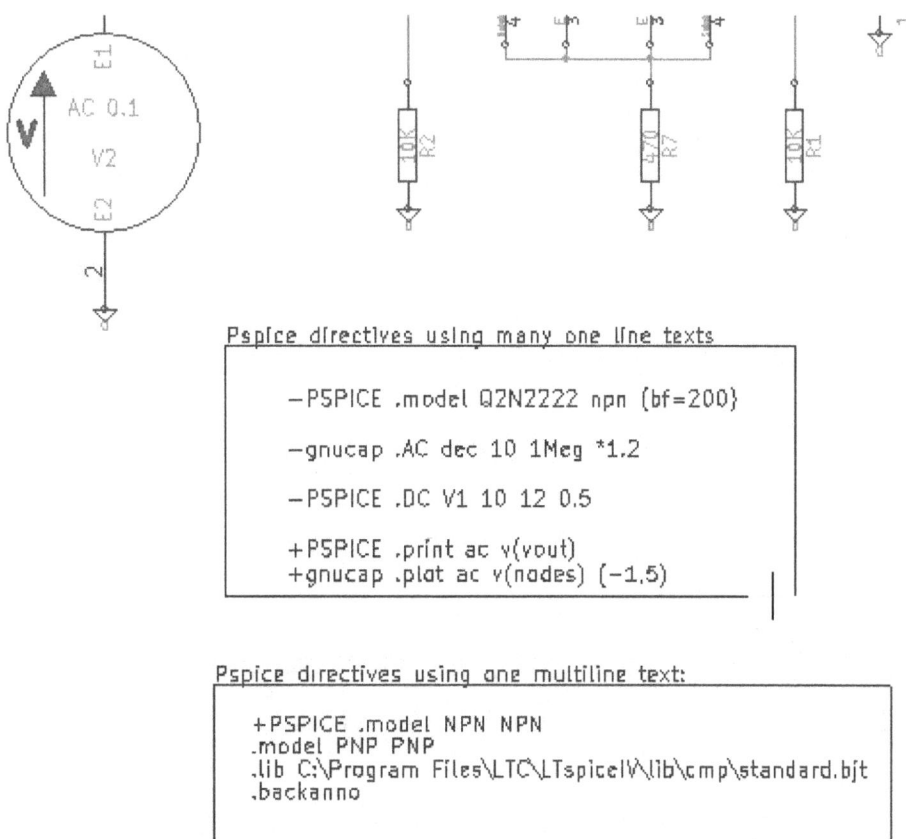

For example: if you type the following text (do not use a label!):

-PSPICE .PROBE

a line .PROBE will be inserted in the netlist.

In the previous example three lines were inserted at the beginning of the netlist and two at the end with this technique.

If you are using multiline texts, **+pspice** or **+gnucap** keywords are needed only once:

+PSPICE .model NPN NPN

.model PNP PNP

.lib C:\Program Files\LTC\LTspiceIV\lib\cmp\standard.bjt

.backanno

creates the four lines:

.model NPN NPN

.model PNP PNP

.lib C:\Program Files\LTC\LTspiceIV\lib\cmp\standard.bjt

.backanno

Also note that the equipotential GND must be named 0 (zero) for Pspice.

# 9.5 - Other formats, using «plugins»

For other netlist formats you can add netlist converters. These converters are automatically launched by Eeschema. Chapter 14 gives some explanations and examples of converters.

A converter is a text file (xsl format) but one can use other languages like Python. When using the xsl format, a tool ( xsltproc.exe or xsltproc ) read the intermediate file created by Eeschema, and the converter file to create the output file. In this case, the converter file (a sheet style) is very small and very easy to write.

## 9.5.1 - Init the dialog window

You can add a new netlist plug-in via the Add Plugin tab.

Here is the plug-in PadsPcb setup window

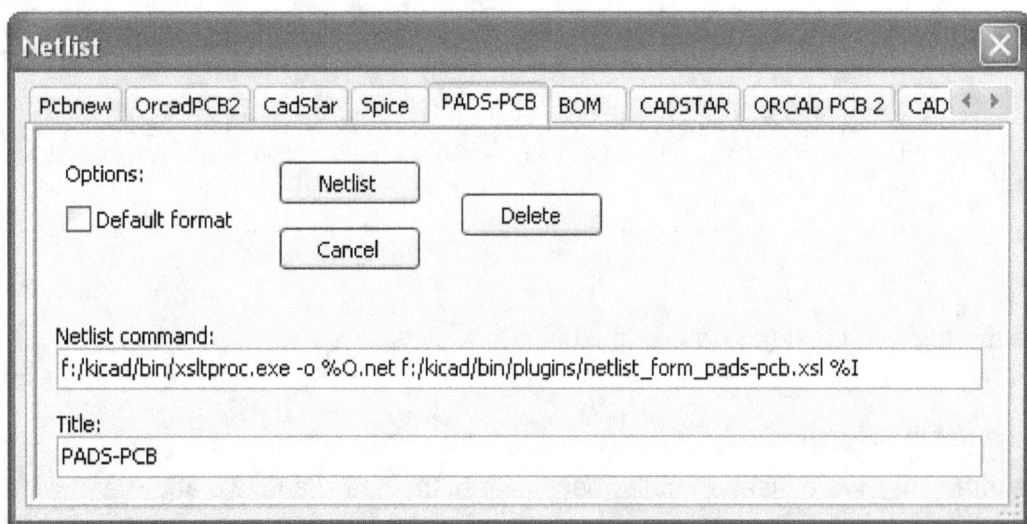

The setup will require:

- A title (for instance: the name of the netlist format).
- The plug-in to launch.

When the netlist is generated:

1. Eeschema creates an intermediate file *.tmp, for instance test.tmp.
2. Eeschema run the plug-in, which reads test.tmp and creates test.net.

### 9.5.2 - Command line format

Here is an example, using xsltproc.exe as tool to convert .xsl files, and a file netlist_form_pads-pcb.xsl as converter sheet style:

**f:/kicad/bin/xsltproc.exe -o %O.net f:/kicad/bin/plugins/netlist_form_pads-pcb.xsl %I**
With:

| f:/kicad/bin/xsltproc.exe | A tool to read and convert xsl file |
|---|---|
| -o %O.net | Output file: %O will define the output file. |
| f:/kicad/bin/plugins/netlist_form_pads-pcb.xsl | File name converter (a sheet style, xsl format). |
| %I | Will be replaced by the intermediate file created by Eeschema (*.tmp). |

For a schematic named test.sch, the actual command line is:

f:/kicad/bin/xsltproc.exe -o test.net f:/kicad/bin/plugins/netlist_form_pads-pcb.xsl test.tmp.

### 9.5.3 - Converter and sheet style (plug in)

This is a very simple piece of software, because its purpose is only to convert an input text file (the intermediate text file) to an other text file. Moreover, from the intermediate text file, you can create a BOM list.

When using  xsltproc as converter tool only the sheet style will be generated.

### 9.5.4 - Intermediate netlist file format

See Chapter 14 for more explanations about xslproc, the descriptions of intermediate file format, and some examples of sheet style for converters.

# 10 - Plot and Print

## Table of Contents

## 10.1 - Introduction

You can access both print and plot commands via the file menu.

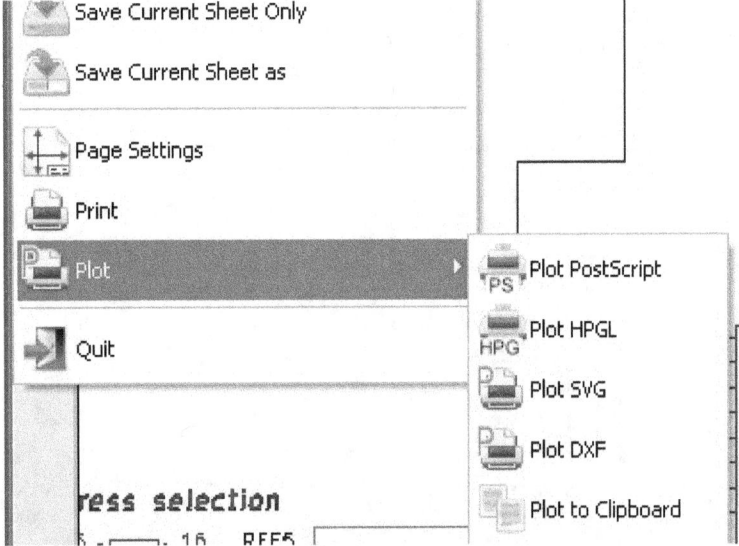

The suported output formats are POSTSCRIPT, HPGL, SVG and DXF. You can as well directly print to your printer.

## 10.2 - Common printing commands

"Plot All" allows you to plot the whole hierarchy (one print file is generated for each sheet).

:Plot Current: prints one file for the current sheet only.

## 10.3 - Plot in HPGL

This command allows you to create an HPGL file. This option is available via the icon . In this format you can define.

- Pen number

- Pen thickness (in 0,001 inch).

- Drawing speed (in cm/S).

- Sheet size.

- Print offsets.

The plotter setup dialog window looks like the following.

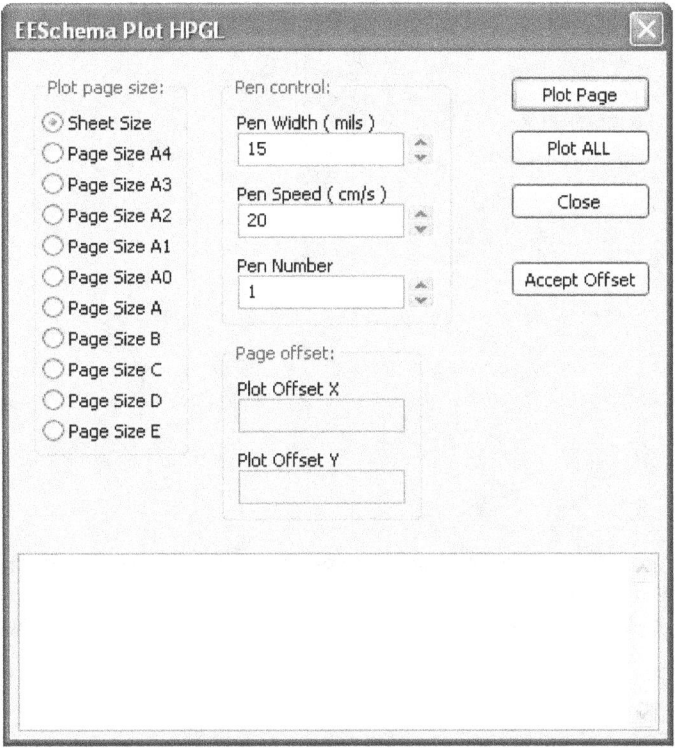

The output file name will be the sheet name plus the extension .plt.

### 10.3.1 - Sheet size selection

Sheet size is normally checked. In this case, the sheet size defined in the title block menu will be used and the chosen scale will be 1. If a different sheet size is selected (A4 with A0, or A with E), the scale is automatically adjusted to fill the page.

### 10.3.2 - Offset adjustments

For all standard dimensions, you can adjust the offsets to center the drawing as accurately as possible. Because plotters have an origin point at the center or at the lower left corner of the sheet, it is necessary to be able to introduce an offset, in order to plot properly.

Generally speaking.

- For plotters having their origin point at the center of the sheet the offset must be negative and set at half of the sheet dimension.
- For plotters having their origin point at the lower left corner of the sheet the offset must be set equal to 0.

To set an offset.

- Select sheet size.
- Set offset X and offset Y.
- Click on accept offset.

## 10.4 - Plot in Postscript

This command allows you to create PostScript files. This option is available via the icon  .

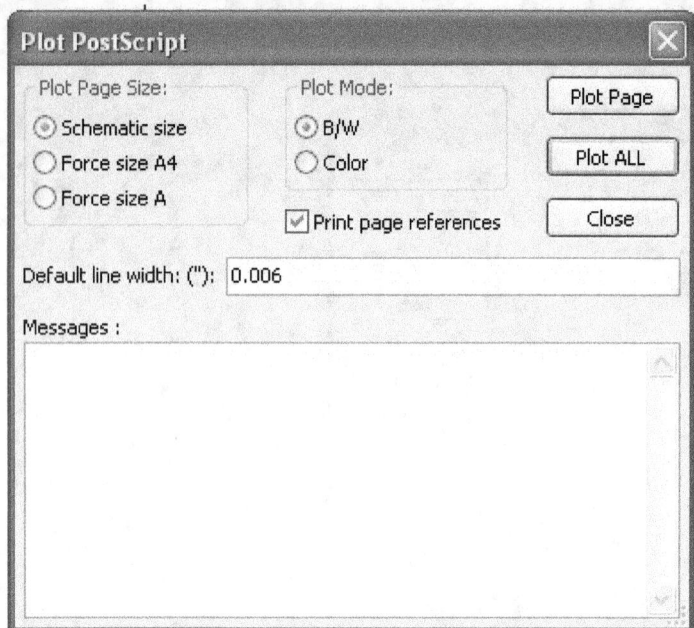

The file name is the sheet name with an extension .ps. You can disable the option "print title block". This is useful if you want to create a postscript file for encapsulation (format .eps) often used to insert a diagram in a word processing software. The message window displays the file names created.

## 10.5 - Plot in SVG

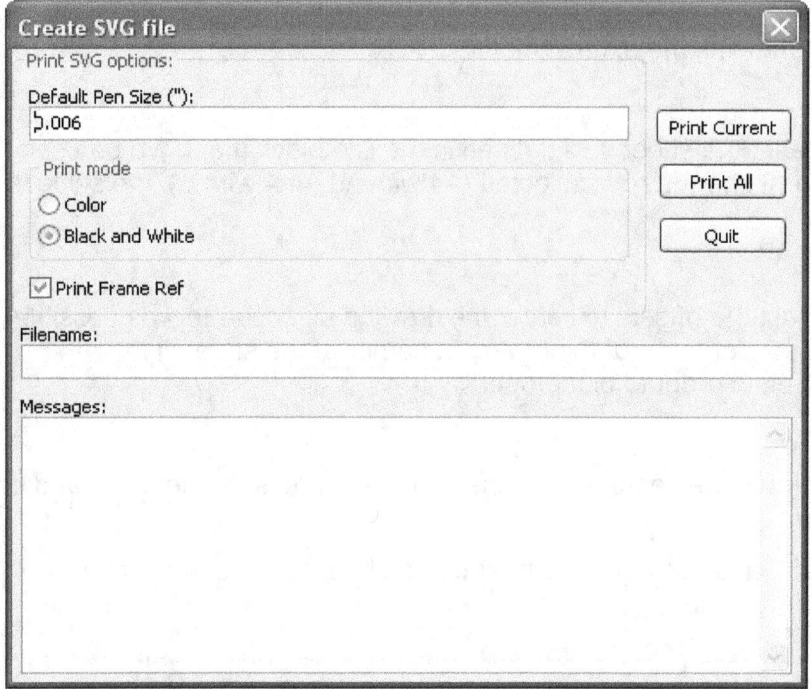

Allows you to create plot files using the vectored format SVG. This option is available via the icon  . The file name is the sheet name with an extension .svg.

## 10.6 - Plot in DXF

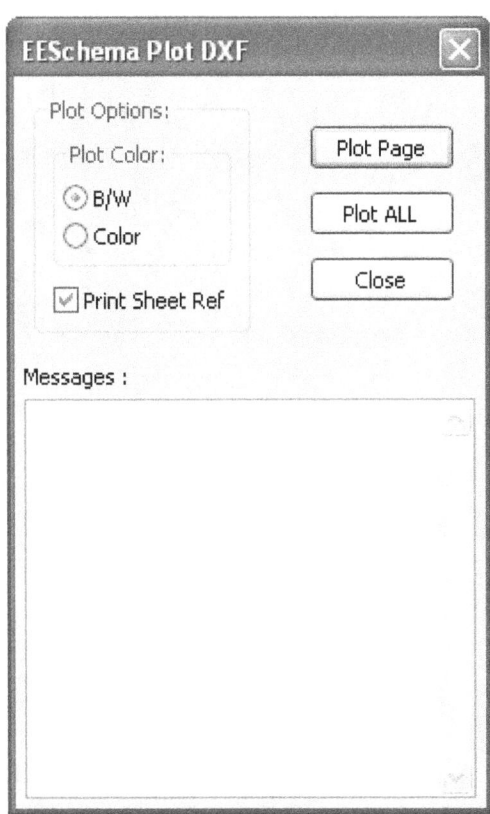

Allows you to create plot files using the format DXF. This option is available via the icon 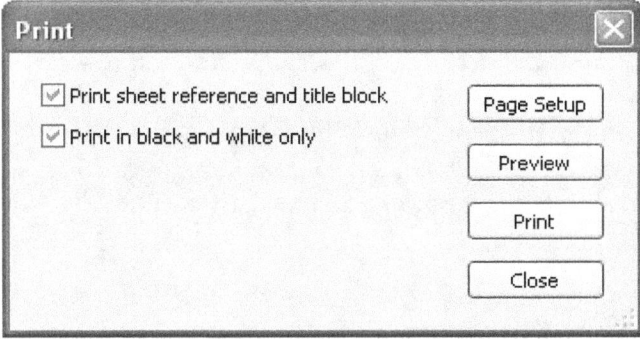. The file name is the sheet name with an extension .dxf.

## 10.7 - Print on paper

This command, available via the icon ![icon], allows you to visualize and generate design files for the standard printer.

The "Print sheet reference and title block" option enables or disables sheet references and title block.

The "Print in black and white" option sets printing in monochrome. This option is generally necessary if you use a black and white laser printer, because colors are printed into half-tones that are often not so readable.

## Table of Contents

## 11.1 - General Information About Component Libraries

A component is a schematic element which contains a graphical representation, electrical connections, and fields defining the component. Components used in a schematic are stored in component libraries. Eeschema provides a component library editing tool that allows you to create libraries, add, delete or transfer components between libraries, export components to files, and import components from files. The library editing tool provides a simple way to manage component library files.

## 11.2 - Component Library Overview

A component library is composed of one or more components. Generally the components are logically grouped by function, type, and/or manufacturer.

A component is composed of:

- Graphical items (lines, circles, arcs, text, etc) that provide the symbolic definition.
- Pins which have both graphic properties (line, clock, inverted, low level active, etc) and electrical properties (input, output, bidirectional, etc.) used by the Electrical Rules Check (ERC) tool.
- Fields such as references, values, corresponding footprint names for PCB design, etc.
- Aliases used to associate a common component such as a 7400 with all of it's derivatives such as 74LS00, 74HC00, and 7437.  All of these aliases share the same library component.

Proper component designing requires:

- Defining if the component is made up of one or more units.
- Defining if the component has an alternate body style also known as a DeMorgan representation.
- Designing it's symbolic representation using lines, rectangles, circles, polygons and text.
- Adding pins by carefully defining each pin's graphical elements, name, number, and electrical properties (input, output, tri-state, power port, etc.).
- Adding an alias if other components have the same symbol and pin out or removing one if the component has been created from an other component.
- Adding optional fields such the name of the footprint used by the PCB design software and/or defining their visibility.
- Documenting the component by adding a description string and links to data sheets, etc.
- Saving it in the desired library.

## 11.3 - Component Library Editor Overview

The component library editor main window is shown below.  It consists three tool bars for quick access to common features and a component viewing/editing area.  Not all commands are available on the tool bars but can be accessed using the menus.

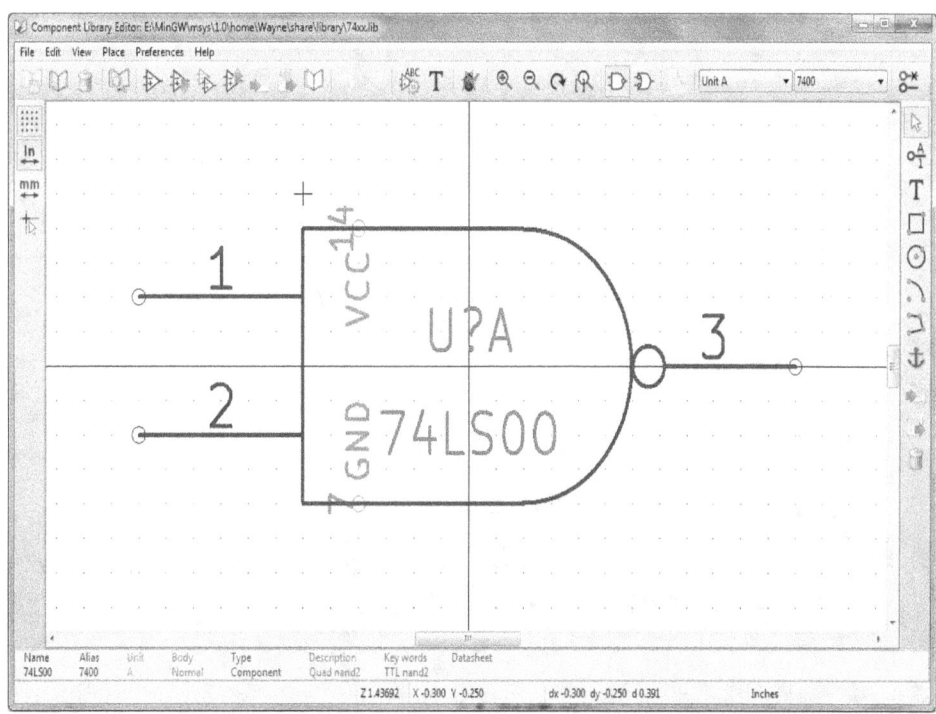

## 11.3.1 - Main Toolbar

The main tool bar typically located at the top of main window shown below consists of the library management tools, undo/redo commands, zoom commands, and component properties dialogs.

| | |
|---|---|
| | Save the currently select library. The button will be disabled if no library is currently selected or no changes to the currently selected library have been made. |
| | Select the library to edit. |
| | Delete a component from the currently selected library or any library defined by the project if no library is currently selected. |
| | Open the component library browser to select the library and component to edit. |
| | Create a new component. |
| | Load component from currently selected library for editing. |
| | Create a new component from the currently loaded component. |
| | Save the current component changes in memory. The library file is not changed. |
| | Import one component from a file. |
| | Export the current component to a file. |
| | Create a new library file containing the current component. Note: new libraries are not automatically added to the project. |
| | Undo last edit. |
| | Redo last undo. |
| | Edit the current component properties. |
| T | Edit the fields of current component. |
| | Test the current component for design errors. |
| | Zoom in. |
| | Zoom out. |
| | Refresh display. |
| | Zoom to fit component in display. |
| | Select the normal body style. The button is disabled if the current component does not have an alternate body style. |

| | |
|---|---|
| | Select the alternate body style. The button is disabled if the current component does not have an alternate body style. |
| | Show the associated documentation. The button will be disabled if no documentation is defined for the current component. |
| Composant A ⌄ | Select the unit to display. The drop down control will be disable if the current component is not derived from multiple units. |
| 7400 ⌄ | Selection the alias. The drop down control will be disabled if the current component does not have any aliases. |
| | Pin editing: independent editing for pin shape and position for components with multiple units and alternate symbols. |

## 11.3.2 - Element Toolbar

The vertical toolbar typically located on the right hand side of the main window allows you to place all of the elements required to design a component. The table below defines each tool bar button.

| | |
|---|---|
| | Select tool. Right clicking with the select tool opens the context menu for the object under the cursor. Left clicking with the select tool displays the attributes of the object under the cursor in the message panel at the bottom of the main window. Left double-click with the select tool will open the properties dialog for the object under the cursor. |
| | Pin tool. Left click to add a new pin. |
| T | Graphical text tool. Left click to add a new graphical text item. |
| | Rectangle tool. Left click to begin drawing the first corner of a graphical rectangle. Left click again to place the opposite corner of the rectangle. |
| | Circle tool. Left click to begin drawing a new graphical circle from the center. Left click again to define the radius of the cicle. |
| | Arc tool. Left click to begin drawing a new graphical arc item from the center. Left click again to define the first arc end point. Left click again to defint the second arc end point. |
| | Polygon tool. Left click to begin drawing a new graphical polygon item in the current component. Left click for each addition polygon line. Left double click to complete the polygon. |
| | Anchor tool. Left click to set the anchor position of the component. |
| | Import a component from a file. |
| | Export the current component to a file. |
| | Delete tool. Left click to delete an object from the current component. |

## 11.3.3 - Options Toolbar

The vertical tool bar typically located on the left hand side of the main window allows you to set some of the editor drawing options. The table below defines each tool bar button.

| | |
|---|---|
| | Toggle grid visibility on and off. |
| In | Set units to inches. |

| mm | Set units to millimeters. |
|---|---|
| ⊤⸝ | Toggle full screen cursor on and off. |

## 11.4 - Library Selection and Maintenance

The selection of the current library is possible via the *select current library icon* which shows you all available libraries and allows you to select one. When a component is loaded or saved, it will be put in this library. The library name of component is the contents of it's value field.

**Note:**

- You must load a library in Eeschema, in order to access it's contents.
- The content of the current library can be saved after modification, by clicking on the *save current library button* on the main tool bar.
- A component can be removed from any library by clicking on the *delete component from library button*.

### 11.4.1 - Select and Save a Component

When you edit a component you are not really working on the component in its library but on a copy of it in the computer's memory. Any edit action can undone easily. A component may be loaded from a local library or from an existing component.

#### 11.4.1.1 - Component Selection

Clicking the *load component from library button* on the main tool bar displays the list of the available components that you can select and load from the currently selected library.

**Note:**

If a component selected by it's alias, the name of the loaded component is displayed on the window title bar instead of selected alias. The list of component aliases is always loaded with each component and can be edited. You can create a new component by selecting an alias of the current component from the *select alias drop down control*. The first item in the alias list is the root name of the component.

**Note:**

Alternatively, clicking the import component button allows you to load a component which has been previously saved by the export component button.

#### 11.4.1.2 - Save a Component

After modification, a component can be saved in the current library or in a new library or exported to a backup file.

To save the modified component in the current library, click the *update changes to library in memory button*. Please note that the update command only saves the component changes in the local memory. This way, you can make up your mind before you save the library.

To permanently save the component changes to the library file, click the *save the current library button* which will overwrite the existing library file with the component changes.

If you want to create a new library containing the current component, click the *save in new library button*. You will be asked to enter a new library name.

**Note:**

New libraries are not automatically added to the current project. You must add any new library you wish to use in a schematic to the list of project libraries in Eeschema using the *component library configuration dialog*.

Click the *export component button* to create a file containing only the current component. This file will be a standard library file which will contains only one component. This file can be used to import the component into another library. In fact the create new library command and the export command are basically identical.

### 11.4.1.3 - Transfer Components to Another Library

You can very easily copy a component from a source library into a destination library using the following commands:

- Select the source library by clicking the *select library button*.
- Load the component to be transferred by clicking the *load component from library button*. The component will be displayed in the editing area.
- Select the destination library by clicking the *select library button*.
- Save the current component to the new library in the local memory by clicking the *update changes to library in memory button*.
- Save the component in the current local library file by clicking the *save the current library button*.

### 11.4.1.4 - Discarding Component Changes

When you are working on a component, the edited component is only a working copy of the actual component in its library. This means that as long as you have not saved it, you can just reload it to discard all changes made. If you have already updated it in the local memory and you have not saved it to the library file, you can always quit and start again. Eeschema will undo all the changes.

## 11.5 - Creating Library Components

### 11.5.1 - Create a New Component

A new component can be created clicking the *new component button*. You will be asked for a component name (this name is used as default value for the value field in the schematic editor), the reference designator (U, IC, R...), the number of units per package (for example a 7400 is made of 4 units per package) and if an alternate body style (sometimes referred to as DeMorgan) is desired. If the reference designator field is left empty, it will default to "U". These properties changed later, but it is preferable to set them correctly at the creation of the component.

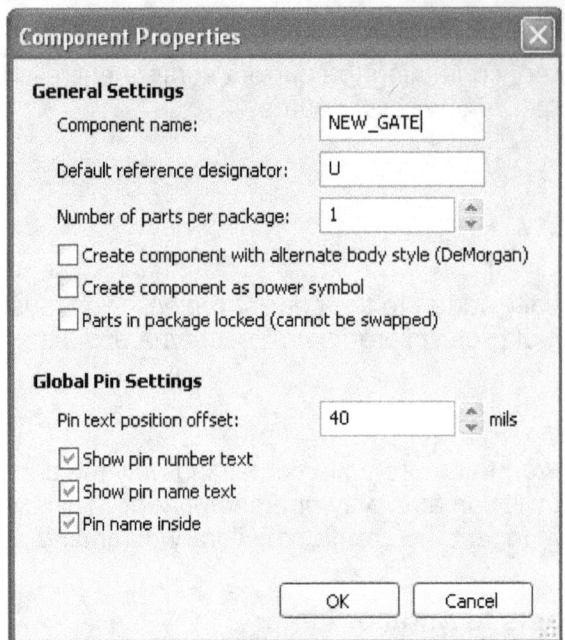

A new component will be created using the properties above and will appear in the editor as shown below.

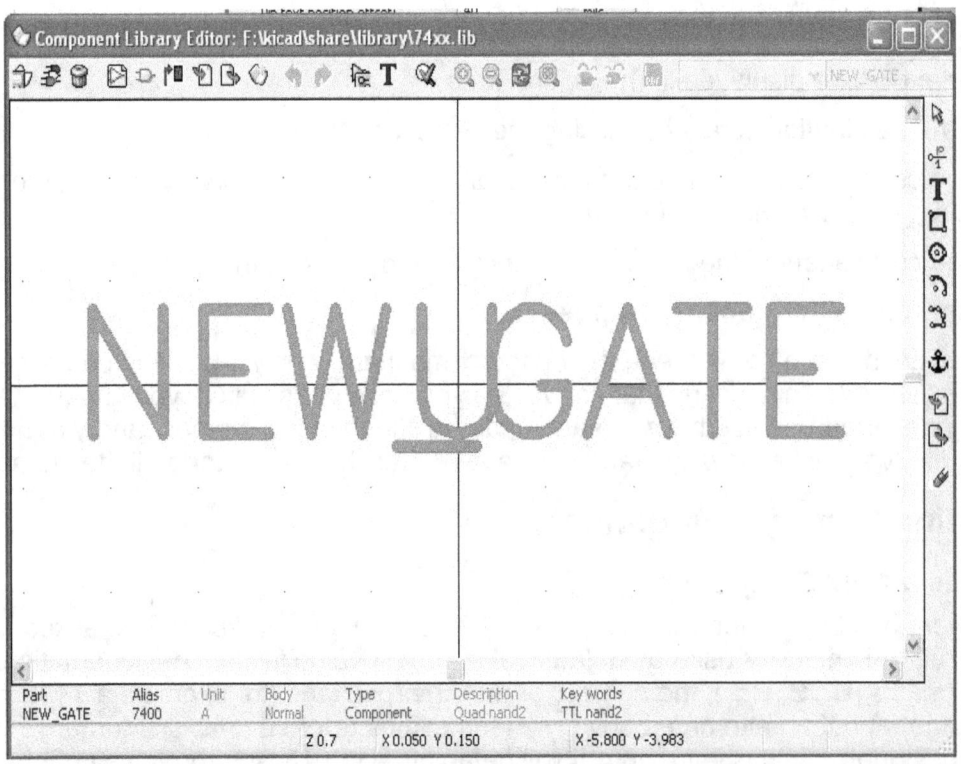

### 11.5.2 - Create a Component from Another Component

Often, the component that you want to make is similar to one already in a component library. In this case it is easy to load and modify an already existing component.

- Load the component which will be used as a starting point.

- Click on the *duplicate component button* or modify its name by right click on the value field and editing the text. If you chose to duplicate the current component, you will be prompted for a new component name.

- If the model component has aliases, you will be prompted to remove aliases from the new component which conflict with the current library. If the answer is no the new component creation will be aborted. Component libraries cannot have any duplicate names or aliases.

- Edit the new component as required.

- Update the new component in the current library by clicking the *update changes to library in memory button* or save to a new library by clicking the *save in new library button* or if you want to save this new component in an other existing library select the other library by clicking on the *select library button* and save the new component.

- Save the current library file to disk by clicking the *save the current library button*.

### 11.5.3 - Component Properties

Component properties should be carefully set during the component creation or alternatively they are inherited from copied component. To change the component properties, click on the *open the component properties* to show the dialog below.

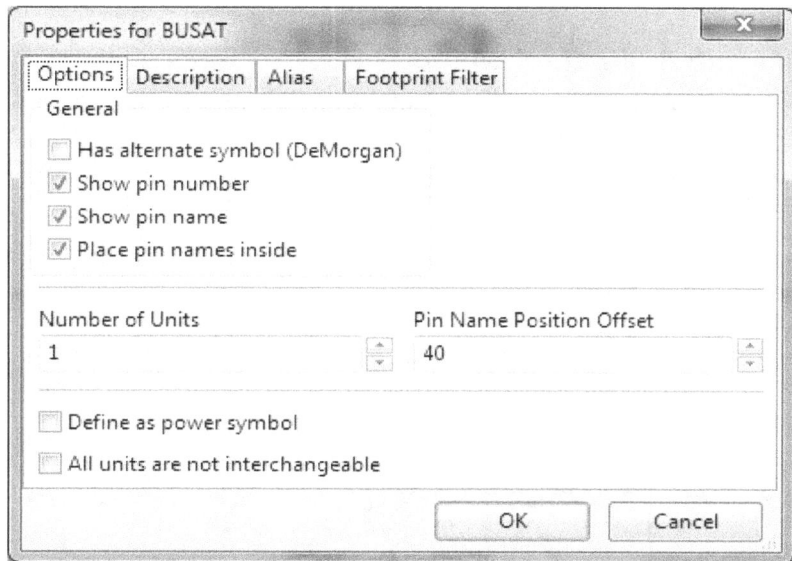

It is very important to correct set the number of units per package and if the component has an alternate symbolic representation parameters correctly because when pins are edited or created the corresponding pins for each unit will created. If you change the number of units per package after pin creation and editing, there will be additional work introduced add the new unit pins and symbols. Nevertheless, it is possible to modify these properies at any time.

The graphic options "Show pin number" and "Show pin name" define the visibility of the pin number and pin name text. This text will be visible if the corresponding options are checked. The option "Place pin names inside" defines the pin name position relative to the pin body. This text will be displayed inside the component outline if the option is checked. In this case the "Pin Name Position Offset" property defines the shift of the text away from the body end of the pin. A value from 30 to 40 (in 1/1000 inch) is reasonable.

The example below shows a component with the "Place pin name inside" option unchecked. Notice the position of the names and pin numbers.

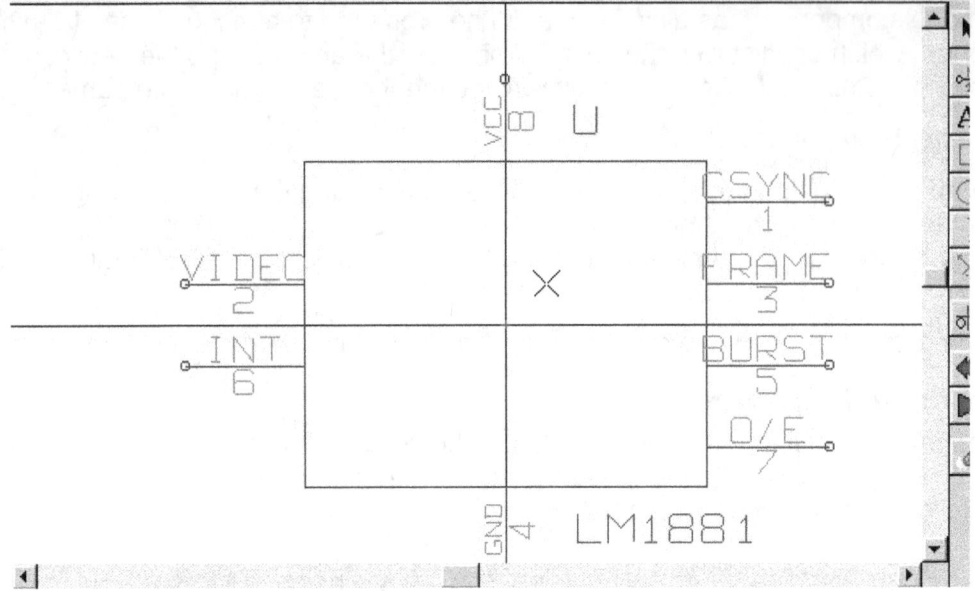

### 11.5.4 - Components with Alternate Symbols

If the component has more than one symbolic repersentation, you will have to select the different symbols of the component in order to edit them. To edit the normal symbol, click the *normal body style button*. To edit the alternate symbol click on the *alternate body style button*. Use the *unit selection drop down control* show below to select the unit you wish to edit.

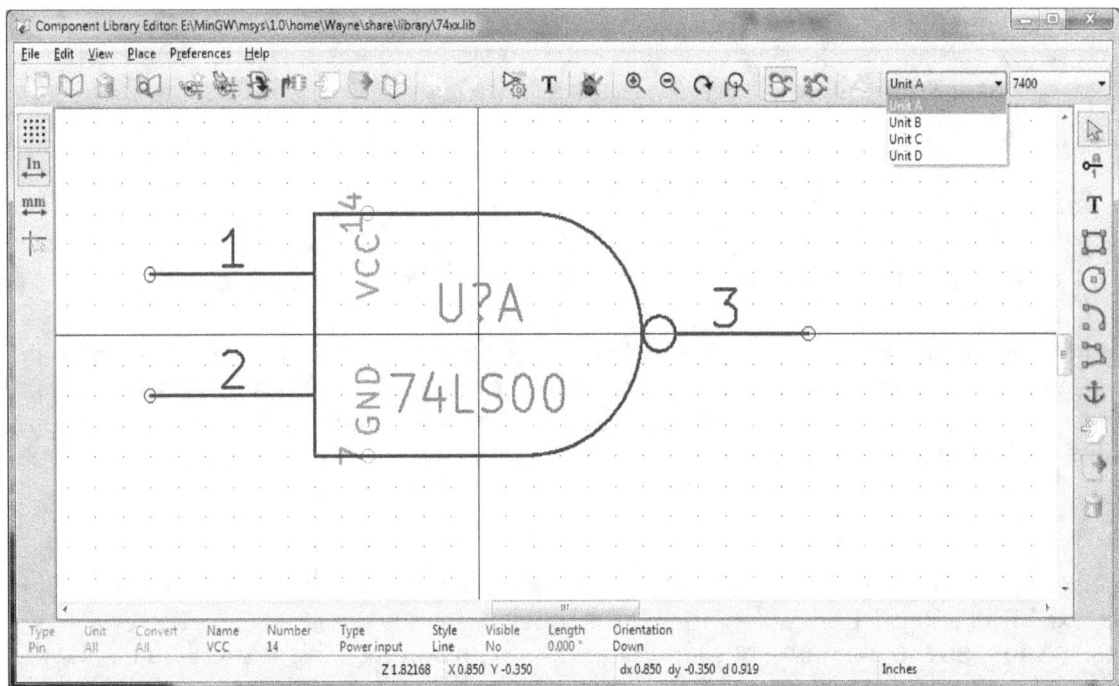

## 11.6 - Graphical Elements

Graphical elements create the symbolic repersentation of a component and contain no electrical connection information. Their design is possible using the following tools:

- Lines and polygons defined by start and end points.
- Rectangles defined by two diagonal corners.
- Circles defined by the center and radius.

- Arcs defined by the starting and ending point of the arc and its center. An arc goes from 0° to 180°.

The vertical toolbar on the right hand side of the main window allows you to place all of the graphical elements required to design a component's symbolic representation.

### 11.6.1 - Graphical Element Membership

Each graphic element (line, arc, circle, etc.) can be defined as common to all units and/or body styles or specific to a given unit and/or body style. Element options can be quickly accessed by the right clicking on the element to display the context menu for the selected element. Below is the context menu for a line element.

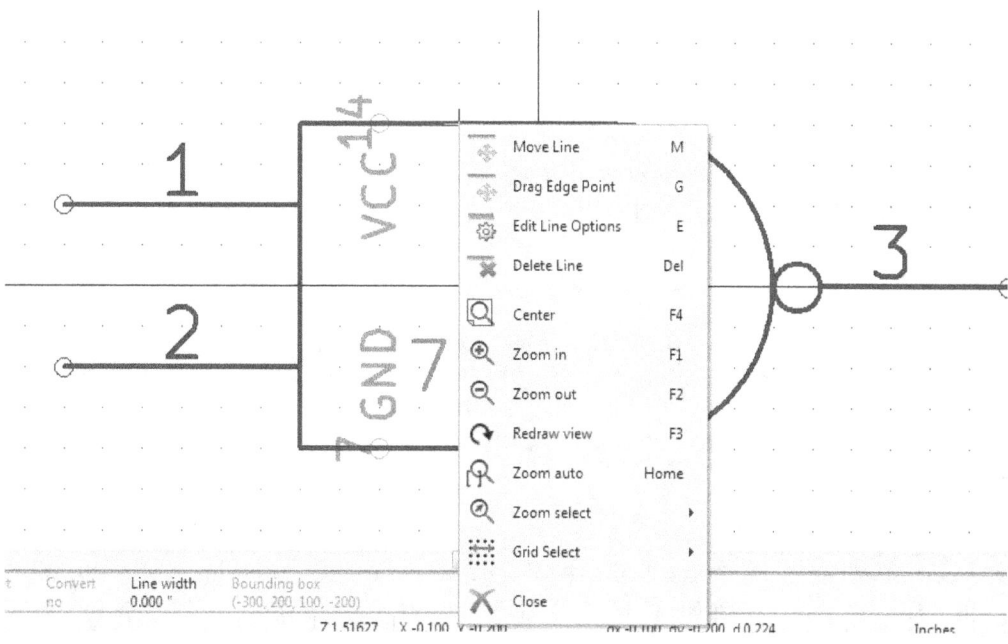

You can also double left click on an element to modify it's properties. Below is the properties dialog for a polygon element.

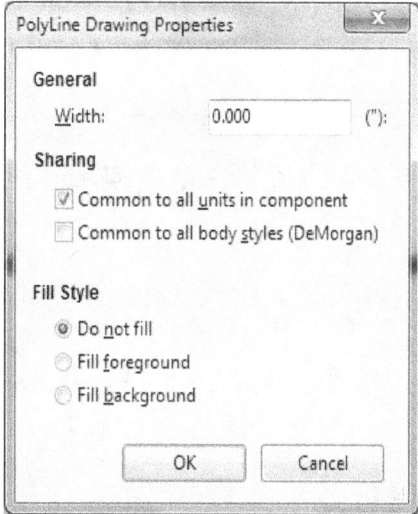

The properties of a graphic element are:

- Line width which defines the width of the element's line in the current drawing units.

- The "Common to all units in component" setting defines if the graphical element is drawn for each unit in component with more than one unit per package or if the graphical element is only drawn for the current unit.
- The "Common by all body styles (DeMorgan)" setting defines if the graphical element is drawn for each symbolic representation in components with an alternate body style or if the graphical element is only drawn for the current body style.
- The fill style setting determines if the symbol defined by the graphical element is to be drawn unfilled, background filled, or foreground filled.

### 11.6.2 - Graphical Text Elements

The *graphical text tool* allows for the creation of graphical text. Graphical text is always readable, even when the component is mirrored. Please note that graphical text items are not fields.

## 11.7 - Multiple Units per Component and Alternate Body Styles

Components can have two symbolic representations ( a standard symbol and an alternate symbol often referred to as "DeMorgan") and/or have more than one unit per package (logic gates for example). Some components can have more than one unit per package each with different symbols and pin configurations.

Consider for instance a relay with two switches which can be designed as a component with three different units: a coil, switch 1, and switch 2. Designing component with multiple units per package and/or alternate body styles is very flexible. A pin or a body symbol item can be common to all units or specific to a given unit or they can be common to both symbolic representations or specific to a given symbol representation.

By default, pins are specific to each symbolic representation of each unit, because the pin number is specific to a unit, and the shape depends on the symbolic representation. When a pin is common to each unit or each symbolic representation, you need to create it only once for all units and all symbolic representations (this is usually the case for power pins). This is also the case for the body style graphic shapes and text, which may be common to each unit. (but typically are specific to each symbolic representation).

### 11.7.1 - Example of a Component Having Multiple Units with Different Symbols:

This is an example of a relay defined with three units per package, switch 1, switch 2, and the coil:

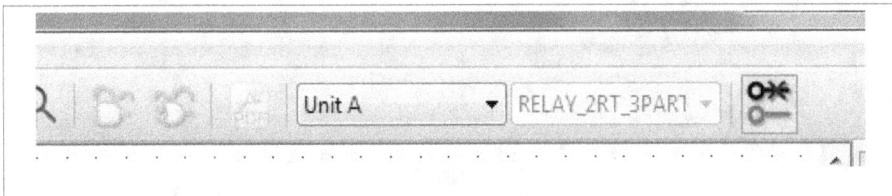

Option: pins are not linked. One can add or edit pins for each unit without any coupling with pins of other units.

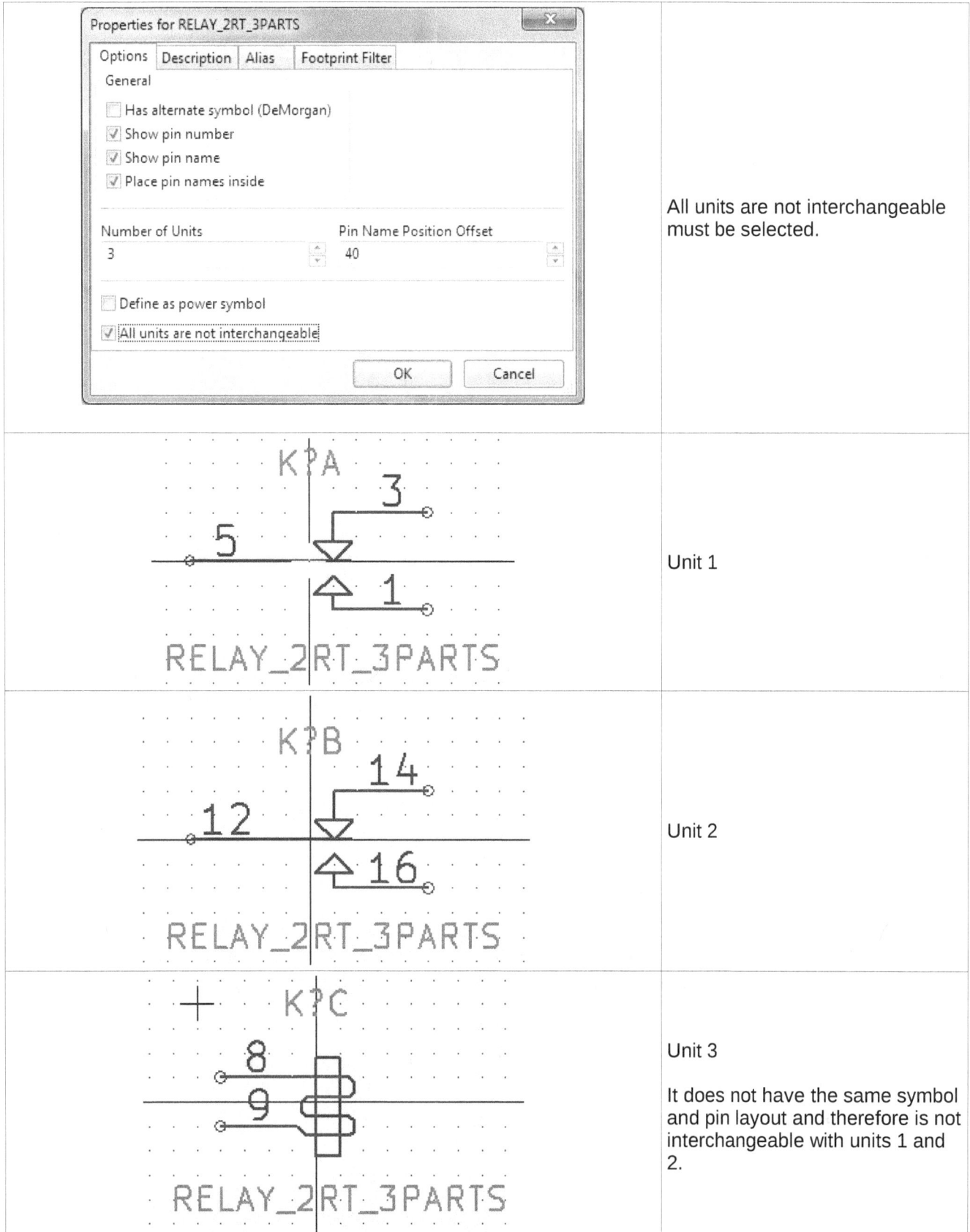

| | |
|---|---|
| | All units are not interchangeable must be selected. |
| | Unit 1 |
| | Unit 2 |
| | Unit 3<br><br>It does not have the same symbol and pin layout and therefore is not interchangeable with units 1 and 2. |

## 11.7.1.1 - Graphical Symbolic Elements

Shown below are properties for a graphic body element. From the relay example above, the three units have different symbolic representations. Therefore, each unit was created separately and the graphical body elements must have the "Common to all units in component" disabled.

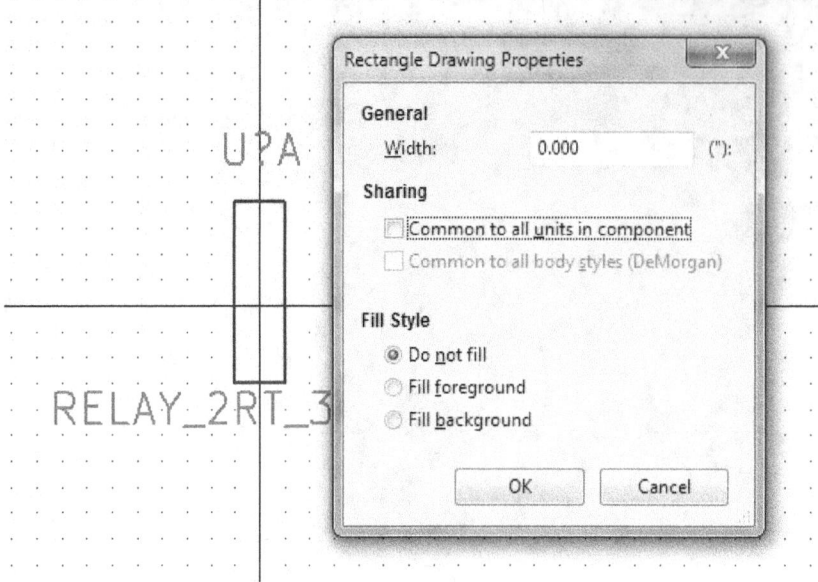

## 11.8 - Pin Creation and Editing

You can click on the *pin tool button* to create and insert a pin.  The editing of all pin properties is done by double-clicking on the pin or right-clicking on the pin to open the pin context menu.  Pins must be created carefully, because any error will have consequences on the PCB design.  Any pin already placed can be edited, deleted, and/or moved.

### 11.8.1 - Pin Overview

A pin is defined by it's graphical representation, it's name and it's "number".  The pin's "number" is defined by a set of 4 letters and/or numbers.  For the electronic rules check (ERC) tool to be useful, the pin's "electrical" type (input, output, tri-state…) must also be defined correctly.  If this type is not defined properly, the schematic ERC check results may be invalid.


Important notes:


- Do not use spaces in pin names and numbers.

- To define a pin name with an inverted signal (overline) use the tilde "~" character.  The next "~" character will turn off the overline.  For example ~FO~O would display $\overline{FO}$O.

- If the pin name is reduced to a single symbol, the pin is regarded as unnamed.

- Pin names starting with "#", are reserved for power port symbols.

- A pin "number" consists of 1 to 4 letters and/or numbers.  1,2,..9999 are valid numbers.  A1, B3, Anod, Gnd, Wire, etc. are also valid.

- Duplicate pin "numbers" cannot exist in a component.

## 11.8.2 - Pin Properties

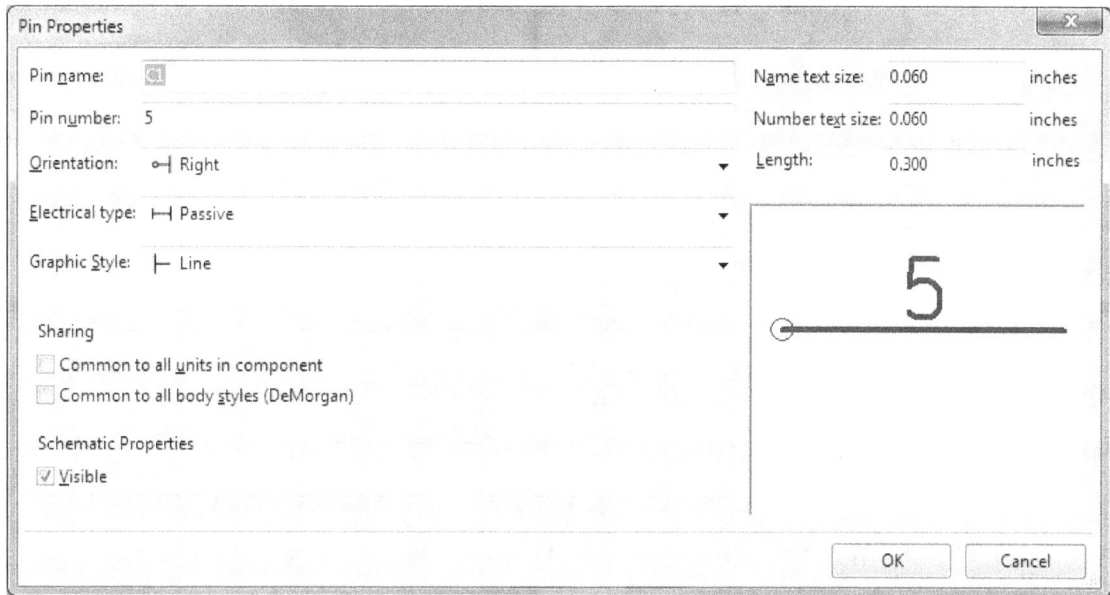

The pin properties dialog allows you to edit all of the characteristics of a pin. This dialog pops up automatically when you create a pin or when double-clicking on an existing pin. This dialog allows you modify:

- Name and name's text size.
- Number and number's text size.
- Length.
- Electrical and graphical types.
- Unit and alternate representation membership.
- Visibility.

## 11.8.3 - Pins Graphical Styles

You can see on the figure below the different pin graphical styles. The choice of graphic styles does not have any influence on the pin's electrical type.

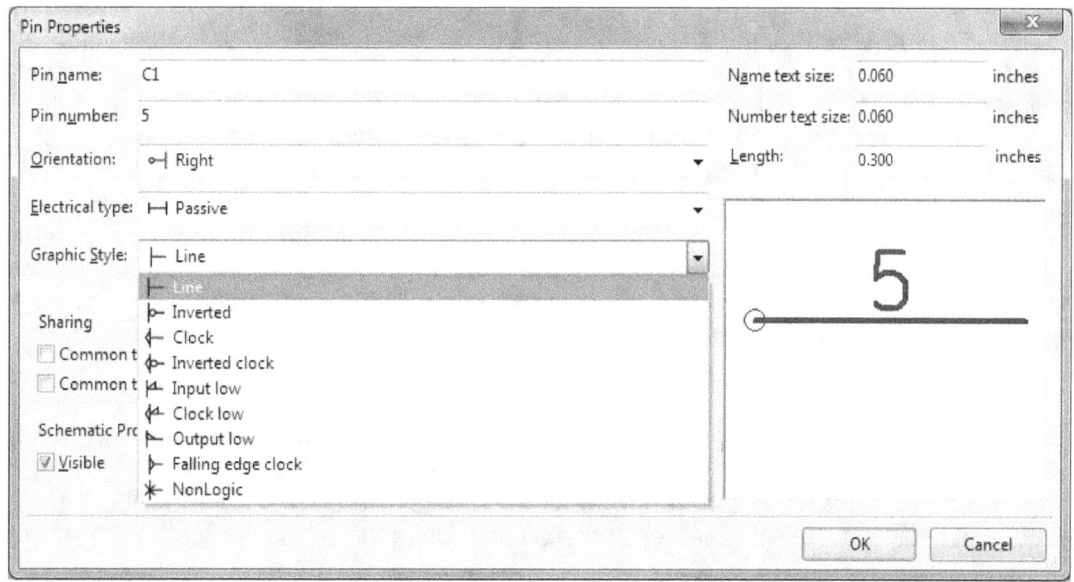

### 11.8.4 - Pin Electrical Types

Choosing the correct electrical type is important for the schematic ERC tool.  The electrical types defined are:

* Bidirectional which indicates bidirectional pins commutable between input and output (microprocessor data bus for example).

* Tri-state is the usual 3 states output.

* Passive is used for passive component pins, resistors, connectors, etc.

* Unspecified can be used when the ERC check doesn't matter.

* Power input is used for the component's power pins. Power pins are automatically connected to the other power input pins with the same name.

* Power output is used for regulator outputs.

* Open emitter and open collector types can be used for logic outputs defined as such.

* Not connected is used when a component has a pin that has no internal connection.

### 11.8.5 - Pin Global Properties

You can modify the length or text size of the name and/or number of all the pins using the Global command entry of the pin context menu.  Click on the parameter you want to modify and type the new value which will then be applied to all of the current component's pins.

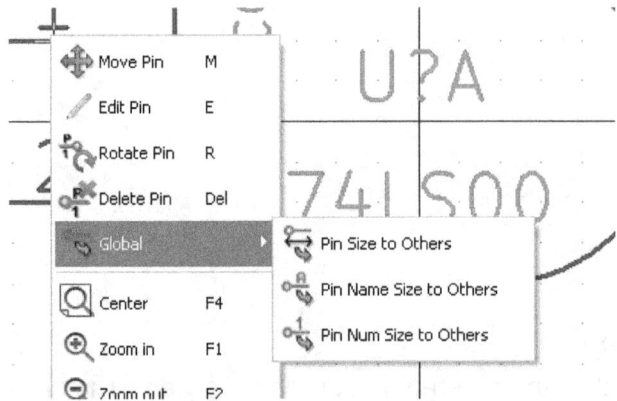

### 11.8.6 - Defining Pins for Multiple Units and Alternate Symbolic Representations

Components with multiple units and/or graphical representations are particularly problematic when creating and editing pins. The majority of pins are specific to each unit (because their pin number is specific to each unit) and to each symbolic representation (because their form and position is specific to each symbolic representation).  The creation and the editing of pins can be problematic for components with multiple units per package and alternate symbolic representations.  The component library editor allows the simultaneous creation of pins.  By default, changes made to a pin are made for all units of a multiple unit component and both representations for components with an alternate representation.  The only exception to this is the pin's graphical type and name.  This dependency was established to allow for easier pin creation and editing in most of the cases.  This dependency can be disabled by toggling the *independent pin edit button* on the main tool bar.  This will allow you to create pins for each unit and representation completely independently.

A component can have two symbolic representations (representation known as "DeMorgan") and can be made up of more than one unit as in the case of components with logic gates.  For certain components, you may want several different graphic elements and pins.  Like the relay sample shown in section 11.7.1, a relay can be represented three distinct units: a coil, switch contact 1, and switch contact 2.

The management of the components with multiple units and components with alternate symbolic representations is flexible. A pin can be common or specific to different units. A pin can also be common to both symbolic representations or specific to each symbolic representation.

By default, pins are specific to each representation of each unit, because their number differs for each unit, and their design is different for each symbolic representation. When a pin is common to all units, it only has to drawn once such as in the case of power pins.

An example is the output pin 7400 quad dual input NAND gate. Since there are four units and two symbolic representations, there are eight separate output pins defined in the component definition. When creating a new 7400 component, unit A of the normal symbolic representation will be shown in the library editor. To edit the pin style in alternate symbolic representation, it must first be enabled by clicking the *show alternate body sytle* button on the tool bar. To edit the pin number for each unit, select the appropriate unit using the *unit selection* drop down control.

## 11.9 - Component Fields

All library components are defined with four default fields. The reference designator, value, footprint assignment, and documentation file link fields are created whenever a component is created or copied. Only the reference designator and value fields are required. For existing fields, you can use the context menu commands by right clicking on the pin. Components defined in libraries typically are defined with these four default fields. Additional fields such as vendor, part number, unit cost, etc. can be added to library components but generally this is done in the schematic editor so the additional fields can be applied to all of the components in the schematic.

### Editing Component Fields

To edit an existing component field, right click on the field text to show the field context menu shown below.

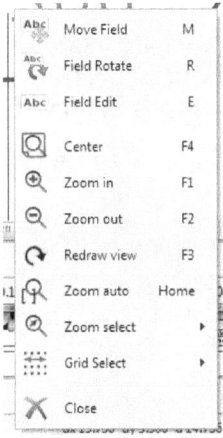

To edit undefined fields, add new fields, or delete optional fields *click the open field properties dialog button* on the main tool bar to open the field properties dialog shown below.

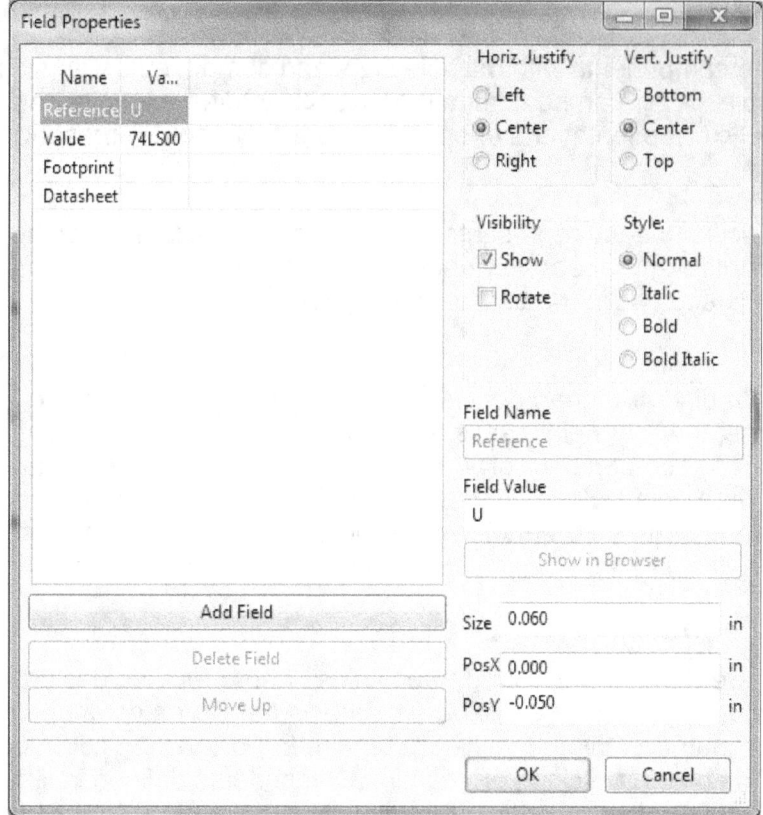

Fields are text sections associated with the component. Do not confused them with the text belonging to the graphic representation of this component.

Important notes:

- Modifying value field effectively creates a new component using using the current component as the starting point for the new component. This new component has the name contained in the value field when you save it to the currently selected library.

- The field edit dialog above must be used to edit a field that is empty or has the invisible attribute enable.

- The footprint is defined as an absolute footprint using the LIBNAME:FPNAME format where LIBNAME is the name of the footprint library defined in the footprint library table (see the "Footprint Library Table" section in the Pcbnew "Reference Manaul") and FPNAME is the name of the footprint in the library LIBNAME.

## 11.10 - Power Symbols

Power symbols are created the same way as normal components. It may be useful to place them in a dedicated library such as power.lib. Power symbols consist of a graphical symbol and a pin of the type "Power Invisible". Power port symbols are handled like any other component by the schematic capture software. Some precautions are essential. Below is an example of a power + 5V symbol.

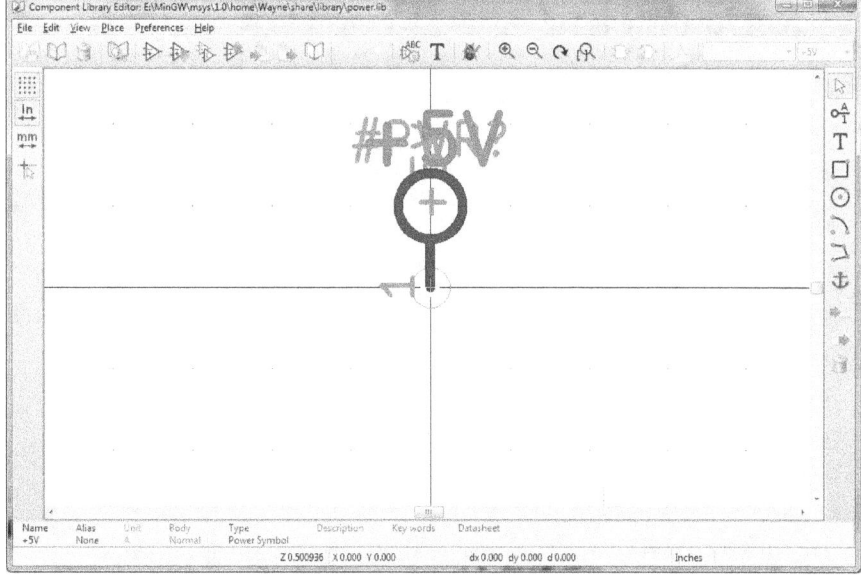

To create a power symbol, use the following steps:

- Add a pin of type "Power input" named + 5V (important because this name will establish connection to the net + 5V), with a pin number of 1 (number of no importance), a length of 0, and a "Line" "Graphic Style".

- Place a small circle and a segment from the pin to the circle as shown.

- The anchor of the symbol is on the pin.

- The component value is +5V.

- The component reference is #+5V. The reference text is no importance except the first character which must be "#" to indicate that the component is a power symbol. By convention, every component in which the reference field starts with a '#' will not appear in the component list or in the netlist and the reference is declared as invisible.

An easier method to create of a new power port symbol is to use another symbol as model.

You just need to:

- Load an existing power symbol.

- Edit the pin name with name of the new power symbol.

- Edit the value field to the same name as the pin, if you want to display the power port value.

- Save the new component.

## Table of Contents

## 12.1 - Overview

A component consist of the following elements

- A graphical representation (geometrical shapes, texts).
- Pins.
- Fields or associated text used by the post processors: netlist, components list.

Two fields are to be initialized: reference and value.
The name of the design associated with the component, and the name of the associated footprint, the other fields are the free fields, they can generally remain empty, and could be filled during schematic capture.

However, managing the documentation associated with any component facilitates the research, use and maintenance of libraries. The associated documentation consists of

- A line of comment.
- A line of key words such as TTL CMOS NAND2, separated by spaces.
- An attached file name (for example an application note or a pdf file). The default directory for attached files:
  kicad/share/library/doc
  If not found:
  kicad/library/doc
  Under linux:
  /usr/local/kicad/share/library/doc
  /usr/share/kicad/library/doc
  /usr/local/share/kicad/library/doc

Key words allow you to selectively search for a component according to various selection criteria. Comments and key words are displayed in various menus, and particularly when you select a component from the library.

The component also has an anchoring point. A rotation or a mirror is made relatively to this anchor point and during a placement this point is used as a reference position. It is thus useful to position this anchor accurately.

A component can have aliases, i.e. equivalent names. This allows you to considerably reduce the number of components that need to be created (for example, a 74LS00 can have aliases such as 74000, 74HC00, 74HCT00...).

Finally, the components are distributed in libraries (classified by topics, or manufacturer) in order to facilitate their management.

## 12.2 - Position a component anchor

The anchor is at the coordinates (0,0) and it is shown by the blue axes displayed on your screen.

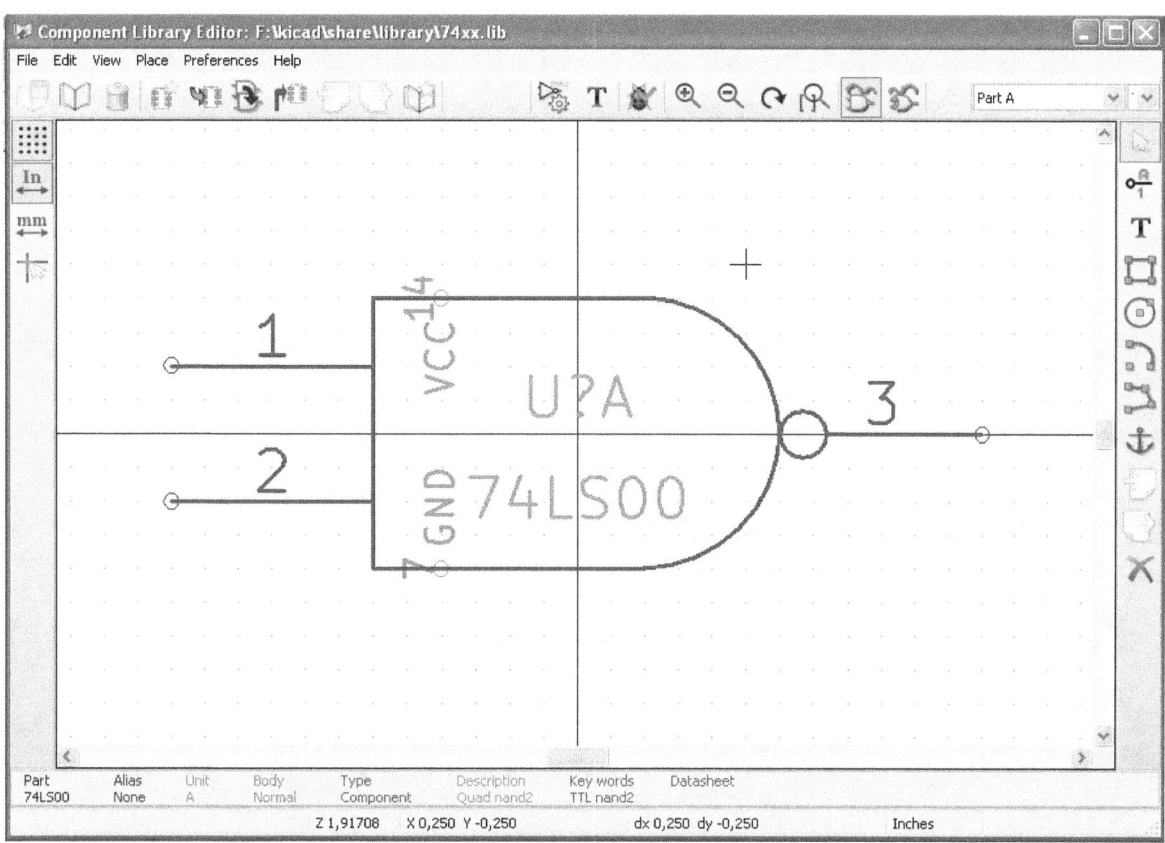

The anchor can be repositioned by selecting the icon ⚓ and clicking on the new desired anchor position. The drawing will be automatically re-centered on the new anchor point.

## 12.3 - Component aliases

An alias is another name corresponding to the same component in the library. Components with similar pin-out and representation can then be represented by only one component, having several aliases (e.g. 7400 with alias 74LS00, 74HC00, 74LS37 ).

The use of aliases allows you to build complete libraries quickly. In addition these libraries, being much more compact, are easily loaded by KiCad.

To modify the list of aliases, you have to select the main editing window via the icon 🖼 and select the alias folder.

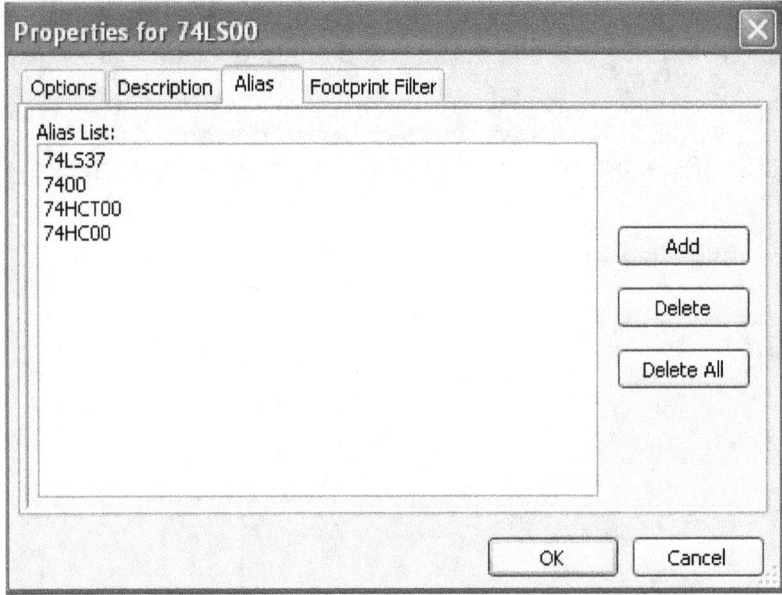

You can thus add or remove the desired alias. The current alias cannot obviously be removed since it is edited.

To remove all aliases, you have firstly to select the root component. The first component in the alias list in the window of selection of the main toolbar.

## 12.4 - Component fields

The field editor is called via the icon $T$ .

There are four special fields (texts attached to the component), and configurable user fields

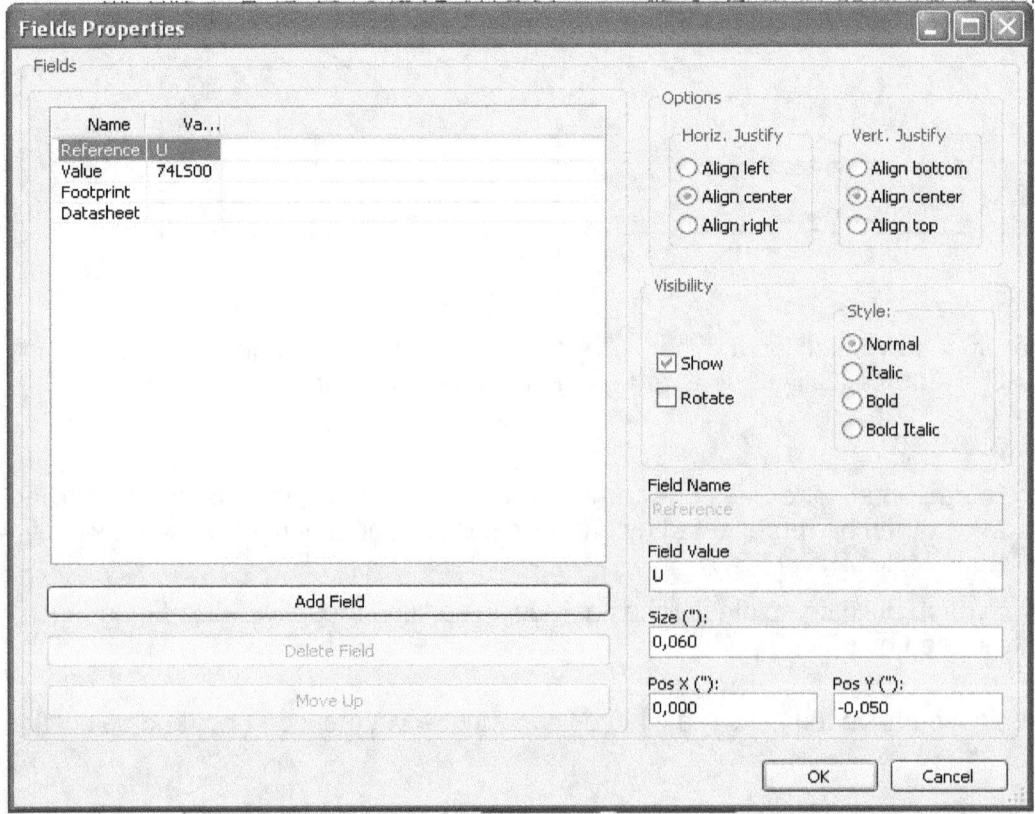

Special fields

- Reference.

- Value. It is the component name in the library and the default value field in schematic.

- Footprint. It is the footprint name used for the board. Not very useful when using CvPcb to setup the footprint list, but mandatory if CvPcb is not used.

- Sheet. It is a reserved field, not used at the time of writing.

## 12.5 - Component documentation

To edit documentation information, it is necessary to call the main editing window of the component via the icon 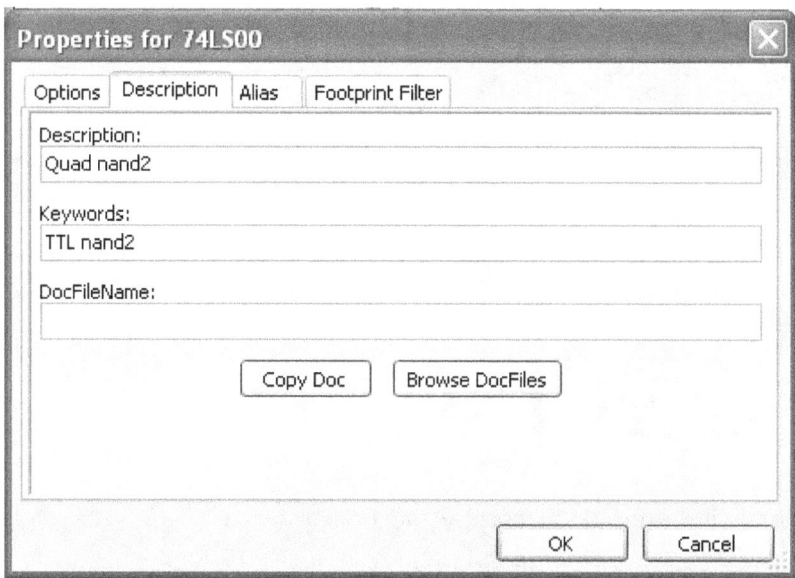 and to select the document folder.

Be sure to select the right alias, or the root component, because this documentation is the only characteristic which differs between aliases. The "Copy Doc" button allows you to copy the documentation information from the root component towards the currently edited alias.

### 12.5.1 - Component keywords

Keywords allow you to search in a selective way for a component according to specific selection criteria (function, technological family, etc.)

The EESchema research tool is not case sensitive. The most current key words used in the libraries are

- CMOS TTL for the logic families

- AND2 NOR3 XOR2 INV... for the gates (AND2 = 2 inputs AND gate, NOR3 = 3 inputs NOR gate).

- JKFF DFF... for JK or D flip-flop.

- ADC, DAC, MUX...

- OpenCol for the gates with open collector output.
  Thus if in the schematic capture software, you search the component: by keys words NAND2 OpenCol EESchema will display the list of components having these 2 key words.

### 12.5.2 - Component documentation (Doc)

The line of comment (and keywords) is displayed in various menus, particularly when you select a component in the displayed components list of a library and in the ViewLib menu.

If this Doc. file exists, it is also accessible in the schematic capture software, in the pop-up menu displayed by right-clicking on the component.

### 12.5.3 - Associated documentation file (DocFileName)

Indicates an attached file (documentation, application schematic) available ( pdf file, schematic diagram, etc.).

### 12.5.4 - Footprint filtering for CvPcb

You can enter a list of allowed footprints for the component. This list acts as a filter used by CvPcb to display only the allowed footprints. A void list does not filter anything.

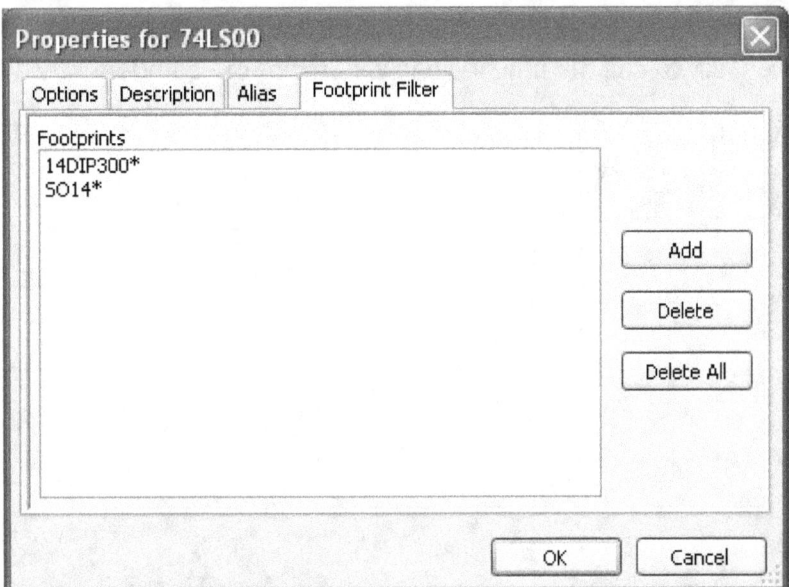

Wild-card characters are allowed.

S014* allows CvPcb to show all the footprints with a name starting by  SO14.

For a resistor, R? shows all the footprints with a 2 letters name starting by R.

Here are samples: with and without filtering

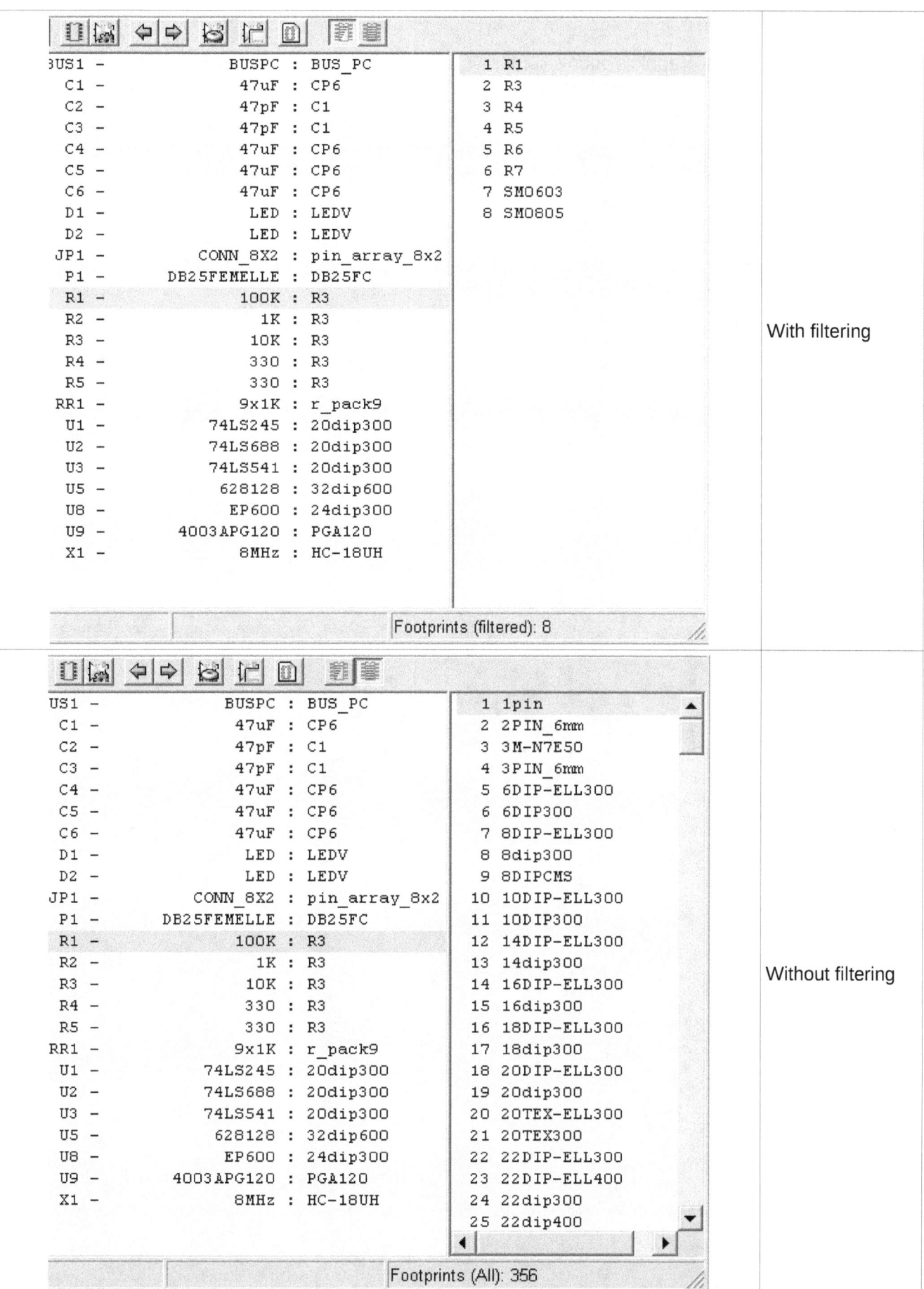

| | | | | | With filtering |
| --- | --- | --- | --- | --- | --- |
| BUS1 – | BUSPC : BUS_PC | | 1 | R1 | |
| C1 – | 47uF : CP6 | | 2 | R3 | |
| C2 – | 47pF : C1 | | 3 | R4 | |
| C3 – | 47pF : C1 | | 4 | R5 | |
| C4 – | 47uF : CP6 | | 5 | R6 | |
| C5 – | 47uF : CP6 | | 6 | R7 | |
| C6 – | 47uF : CP6 | | 7 | SM0603 | |
| D1 – | LED : LEDV | | 8 | SM0805 | |
| D2 – | LED : LEDV | | | | |
| JP1 – | CONN_8X2 : pin_array_8x2 | | | | |
| P1 – | DB25FEMELLE : DB25FC | | | | |
| R1 – | 100K : R3 | | | | |
| R2 – | 1K : R3 | | | | |
| R3 – | 10K : R3 | | | | |
| R4 – | 330 : R3 | | | | |
| R5 – | 330 : R3 | | | | |
| RR1 – | 9x1K : r_pack9 | | | | |
| U1 – | 74LS245 : 20dip300 | | | | |
| U2 – | 74LS688 : 20dip300 | | | | |
| U3 – | 74LS541 : 20dip300 | | | | |
| U5 – | 628128 : 32dip600 | | | | |
| U8 – | EP600 : 24dip300 | | | | |
| U9 – | 4003APG120 : PGA120 | | | | |
| X1 – | 8MHz : HC-18UH | | | | |

Footprints (filtered): 8

Footprints (All): 356

Without filtering

## 12.6 - Symbol library

You can easily compile a graphic symbols library file containing frequently used symbols .This can be used for the creation of components (triangles, the shape of AND, OR, Exclusive OR gates, etc.) for saving and subsequent re-use.

These files are stored by default in the library directory and have a .sym extension. The symbols are not gathered in libraries like the components because they are generally not so many.

### 12.6.1 - Export or create a symbol

A component can be exported as a symbol with the button ![button]. You can generally create only one graphic, also it will be a good idea to delete all pins, if they exist.

### 12.6.2 - Import a symbol

Importing allows you to add graphics to a component you are editing. A symbol is imported with the button ![button]. Imported graphics are added as they were created in existing graphics.

## Table of Contents

## 13.1 - Introduction

Viewlib allows you to quickly examine the content of libraries. Viewlib is called by the tool or by the "place component" tool available from the right-hand side toolbar.

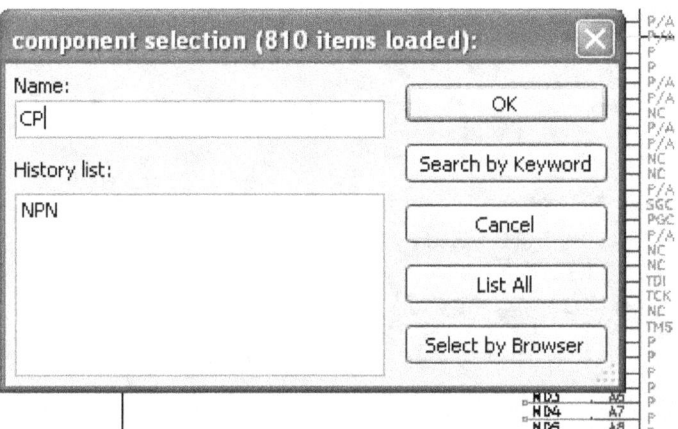

## 13.2 - Viewlib - main screen

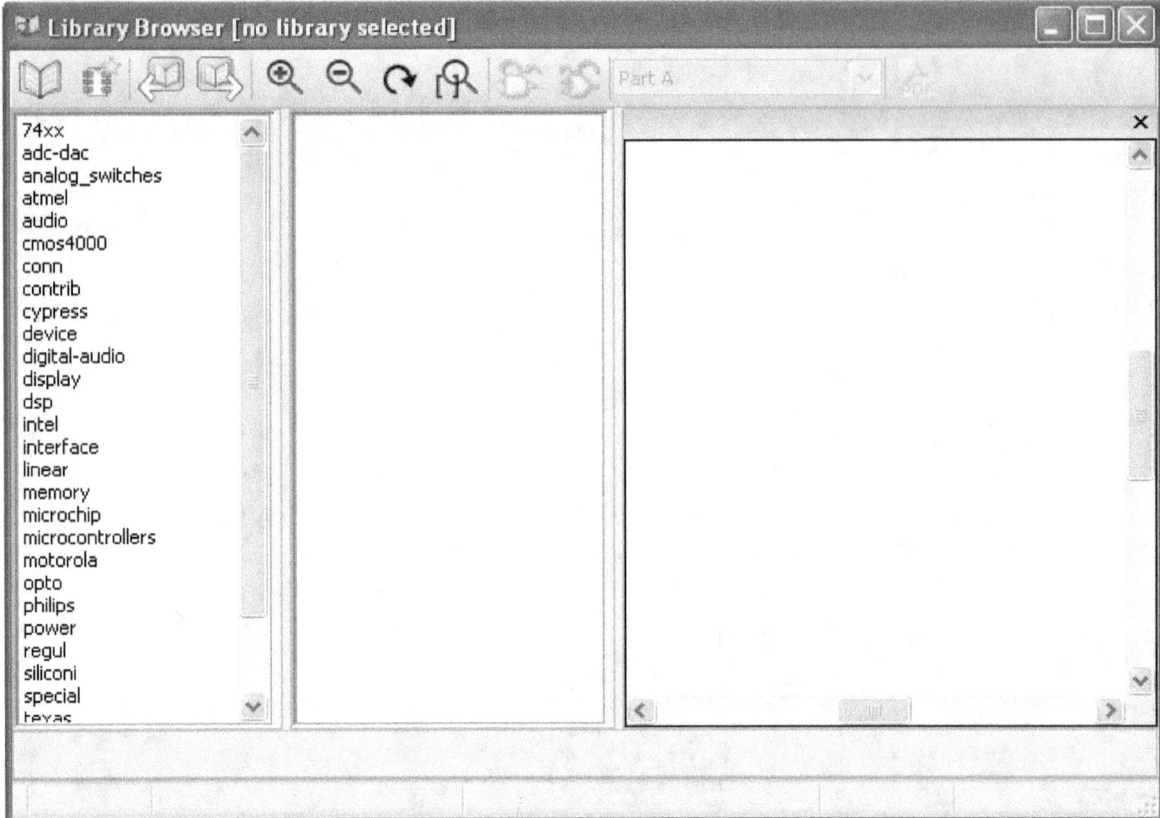

To examine the library content you need to select the wanted library from the list on the left-hand side. Available components will then appear in the second list which allow you to select a component.

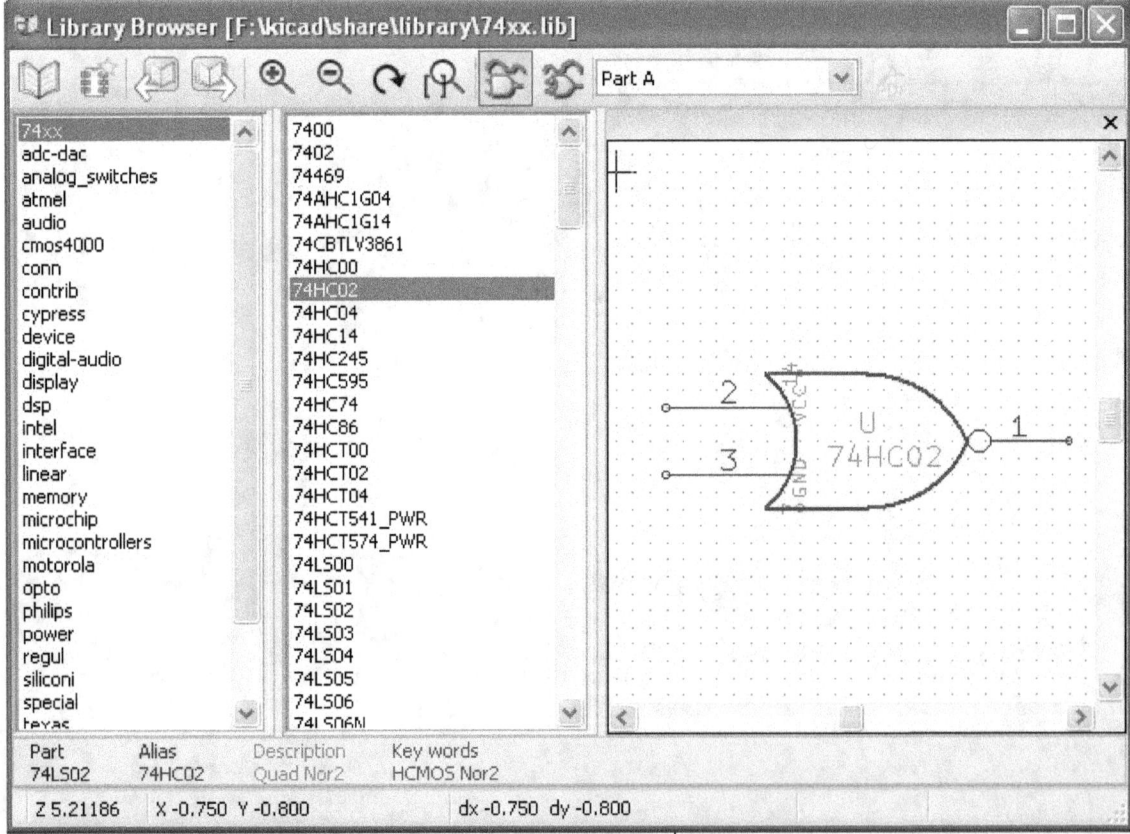

## 13.3 - Viewlib top toolbar

The top tool bar in Viewlib is shown below.

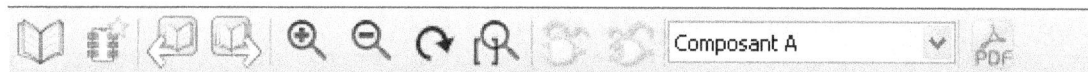

or, when called by the place component dialog frame from Eeschema it appears as below.

The available commands are.

| | |
|---|---|
| | Selection of the desired library which can be also selected in the displayed list. |
| | Selection of the component which can be also selected in the displayed list. |
| | Display previous component. |
| | Display next component. |
| | Zoom tools. |
| | Selection of the representation (normal or converted) if exist. |
| Composant A | Selection of the part (if multi-part component). |
| | If it exist, display the associated documents.<br><br>Exists only when called by the place component dialog frame from Eeschema. |
| | Close Viewlib and place the selected component in Eeschema. |

## Table of Contents

## 14.1 - Intermediate Netlist File

BOM files and netlist files can be converted from an Intermediate netlist file created by Eeschema.

This file uses XML syntax and is called the intermediate netlist. The intermediate netlist includes a large amount of data about your board and because of this, it can be used with post-processing to create a BOM or other reports.

Depending on the output (BOM or netlist), different subsets of the complete Intermediate Netlist file will be used in the post-processing.

## 14.1.1 - Schematic sample

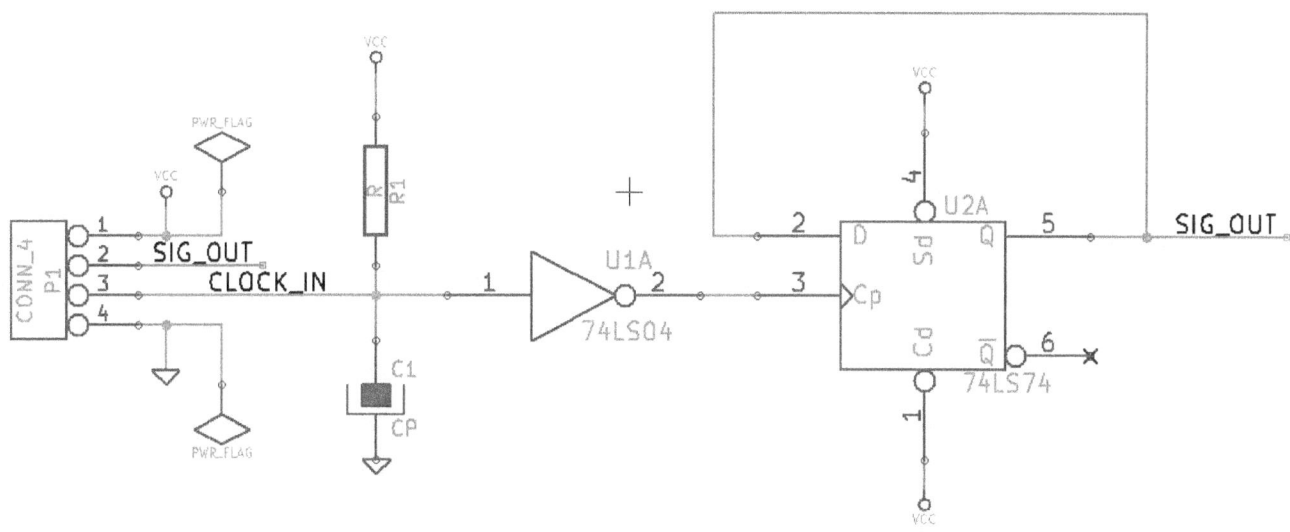

## 14.1.2 - The Intermediate Netlist file sample

The corresponding intermediate netlist (using XML syntax) of the circuit above is shown below.

```
<?xml version="1.0" encoding="utf-8"?>
<export version="D">
  <design>
    <source>F:\kicad_aux\netlist_test\netlist_test.sch</source>
    <date>29/08/2010 20:35:21</date>
    <tool>eeschema (2010-08-28 BZR 2458)-unstable</tool>
  </design>
  <components>
    <comp ref="P1">
      <value>CONN_4</value>
      <libsource lib="conn" part="CONN_4"/>
      <sheetpath names="/" tstamps="/"/>
      <tstamp>4C6E2141</tstamp>
    </comp>
    <comp ref="U2">
      <value>74LS74</value>
      <libsource lib="74xx" part="74LS74"/>
      <sheetpath names="/" tstamps="/"/>
      <tstamp>4C6E20BA</tstamp>
    </comp>
    <comp ref="U1">
      <value>74LS04</value>
      <libsource lib="74xx" part="74LS04"/>
      <sheetpath names="/" tstamps="/"/>
      <tstamp>4C6E20A6</tstamp>
    </comp>
    <comp ref="C1">
      <value>CP</value>
      <libsource lib="device" part="CP"/>
      <sheetpath names="/" tstamps="/"/>
      <tstamp>4C6E2094</tstamp>
    </comp>
    <comp ref="R1">
      <value>R</value>
      <libsource lib="device" part="R"/>
      <sheetpath names="/" tstamps="/"/>
      <tstamp>4C6E208A</tstamp>
```

```xml
      </comp>
  </components>
  <libparts>
    <libpart lib="device" part="C">
      <description>Condensateur non polarise</description>
      <footprints>
        <fp>SM*</fp>
        <fp>C?</fp>
        <fp>C1-1</fp>
      </footprints>
      <fields>
        <field name="Reference">C</field>
        <field name="Value">C</field>
      </fields>
      <pins>
        <pin num="1" name="~" type="passive"/>
        <pin num="2" name="~" type="passive"/>
      </pins>
    </libpart>
    <libpart lib="device" part="R">
      <description>Resistance</description>
      <footprints>
        <fp>R?</fp>
        <fp>SM0603</fp>
        <fp>SM0805</fp>
        <fp>R?-*</fp>
        <fp>SM1206</fp>
      </footprints>
      <fields>
        <field name="Reference">R</field>
        <field name="Value">R</field>
      </fields>
      <pins>
        <pin num="1" name="~" type="passive"/>
        <pin num="2" name="~" type="passive"/>
      </pins>
    </libpart>
    <libpart lib="conn" part="CONN_4">
      <description>Symbole general de connecteur</description>
      <fields>
        <field name="Reference">P</field>
        <field name="Value">CONN_4</field>
      </fields>
      <pins>
        <pin num="1" name="P1" type="passive"/>
        <pin num="2" name="P2" type="passive"/>
        <pin num="3" name="P3" type="passive"/>
        <pin num="4" name="P4" type="passive"/>
      </pins>
    </libpart>
    <libpart lib="74xx" part="74LS04">
      <description>Hex Inverseur</description>
      <fields>
        <field name="Reference">U</field>
        <field name="Value">74LS04</field>
      </fields>
      <pins>
        <pin num="1" name="~" type="input"/>
        <pin num="2" name="~" type="output"/>
        <pin num="3" name="~" type="input"/>
        <pin num="4" name="~" type="output"/>
        <pin num="5" name="~" type="input"/>
        <pin num="6" name="~" type="output"/>
        <pin num="7" name="GND" type="power_in"/>
        <pin num="8" name="~" type="output"/>
        <pin num="9" name="~" type="input"/>
        <pin num="10" name="~" type="output"/>
        <pin num="11" name="~" type="input"/>
        <pin num="12" name="~" type="output"/>
        <pin num="13" name="~" type="input"/>
        <pin num="14" name="VCC" type="power_in"/>
      </pins>
    </libpart>
    <libpart lib="74xx" part="74LS74">
      <description>Dual D FlipFlop, Set &amp; Reset</description>
```

```
        <docs>74xx/74hc_hct74.pdf</docs>
        <fields>
          <field name="Reference">U</field>
          <field name="Value">74LS74</field>
        </fields>
        <pins>
          <pin num="1" name="Cd" type="input"/>
          <pin num="2" name="D" type="input"/>
          <pin num="3" name="Cp" type="input"/>
          <pin num="4" name="Sd" type="input"/>
          <pin num="5" name="Q" type="output"/>
          <pin num="6" name="~Q" type="output"/>
          <pin num="7" name="GND" type="power_in"/>
          <pin num="8" name="~Q" type="output"/>
          <pin num="9" name="Q" type="output"/>
          <pin num="10" name="Sd" type="input"/>
          <pin num="11" name="Cp" type="input"/>
          <pin num="12" name="D" type="input"/>
          <pin num="13" name="Cd" type="input"/>
          <pin num="14" name="VCC" type="power_in"/>
        </pins>
      </libpart>
    </libparts>
    <libraries>
      <library logical="device">
        <uri>F:\kicad\share\library\device.lib</uri>
      </library>
      <library logical="conn">
        <uri>F:\kicad\share\library\conn.lib</uri>
      </library>
      <library logical="74xx">
        <uri>F:\kicad\share\library\74xx.lib</uri>
      </library>
    </libraries>
    <nets>
      <net code="1" name="GND">
        <node ref="U1" pin="7"/>
        <node ref="C1" pin="2"/>
        <node ref="U2" pin="7"/>
        <node ref="P1" pin="4"/>
      </net>
      <net code="2" name="VCC">
        <node ref="R1" pin="1"/>
        <node ref="U1" pin="14"/>
        <node ref="U2" pin="4"/>
        <node ref="U2" pin="1"/>
        <node ref="U2" pin="14"/>
        <node ref="P1" pin="1"/>
      </net>
      <net code="3" name="">
        <node ref="U2" pin="6"/>
      </net>
      <net code="4" name="">
        <node ref="U1" pin="2"/>
        <node ref="U2" pin="3"/>
      </net>
      <net code="5" name="/SIG_OUT">
        <node ref="P1" pin="2"/>
        <node ref="U2" pin="5"/>
        <node ref="U2" pin="2"/>
      </net>
      <net code="6" name="/CLOCK_IN">
        <node ref="R1" pin="2"/>
        <node ref="C1" pin="1"/>
        <node ref="U1" pin="1"/>
        <node ref="P1" pin="3"/>
      </net>
    </nets>
</export>
```

## 14.2 - Conversion to a new netlist format

By applying a post-processing filter to the Intermediate netlist file you can generate foreign netlist files as well as BOM files. Because this conversion is a text to text transformation, this post-processing filter can be written using Python, XSLT, or any other tool capable of taking XML as input.

XSLT itself is a an XML language very suitable for XML transformations. There is a free program called *xsltproc* that you can download and install.  The xsltproc program can be used to read the Intermediate XML netlist input file, apply a style-sheet to transform the input, and save the results in an output file. Use of xsltproc requires a style-sheet file using XSLT conventions. The full conversion process is handled by Eeschema, after it is configured once to run xsltproc in a specific way.

## 14.3 - XSLT approach

The document that describes XSL Transformations (XSLT) is available here:

```
http://www.w3.org/TR/xslt
```

### 14.3.1 - Create a Pads-Pcb netlist file

The pads-pcb format is comprised of two sections.

* The footprint list.

* The Nets list: grouping pads references by nets.

Immediately below is a style-sheet which converts the Intermediate Netlist file to a pads-pcb netlist format:

```xml
<?xml version="1.0" encoding="ISO-8859-1"?>
<!--XSL style sheet to EESCHEMA Generic Netlist Format to PADS netlist format
    Copyright (C) 2010, SoftPLC Corporation.
    GPL v2.

    How to use:
        https://lists.launchpad.net/kicad-developers/msg05157.html
-->

<!DOCTYPE xsl:stylesheet [
  <!ENTITY nl  "&#xd;&#xa;"> <!--new line CR, LF -->
]>

<xsl:stylesheet version="1.0" xmlns:xsl="http://www.w3.org/1999/XSL/Transform">
<xsl:output method="text" omit-xml-declaration="yes" indent="no"/>

<xsl:template match="/export">
    <xsl:text>*PADS-PCB*&nl;*PART*&nl;</xsl:text>
    <xsl:apply-templates select="components/comp"/>
    <xsl:text>&nl;*NET*&nl;</xsl:text>
    <xsl:apply-templates select="nets/net"/>
    <xsl:text>*END*&nl;</xsl:text>
</xsl:template>

<!-- for each component -->
<xsl:template match="comp">
    <xsl:text> </xsl:text>
    <xsl:value-of select="@ref"/>
    <xsl:text> </xsl:text>
    <xsl:choose>
        <xsl:when test = "footprint != '' ">
            <xsl:apply-templates select="footprint"/>
        </xsl:when>
        <xsl:otherwise>
            <xsl:text>unknown</xsl:text>
        </xsl:otherwise>
    </xsl:choose>
    <xsl:text>&nl;</xsl:text>
</xsl:template>

<!-- for each net -->
<xsl:template match="net">
    <!-- nets are output only if there is more than one pin in net -->
    <xsl:if test="count(node)>1">
```

```
        <xsl:text>*SIGNAL* </xsl:text>
        <xsl:choose>
            <xsl:when test = "@name != '' ">
                <xsl:value-of select="@name"/>
            </xsl:when>
            <xsl:otherwise>
                <xsl:text>N-</xsl:text>
                <xsl:value-of select="@code"/>
            </xsl:otherwise>
        </xsl:choose>
        <xsl:text>&nl;</xsl:text>
        <xsl:apply-templates select="node"/>
    </xsl:if>
</xsl:template>

<!-- for each node -->
<xsl:template match="node">
    <xsl:text> </xsl:text>
    <xsl:value-of select="@ref"/>
    <xsl:text>.</xsl:text>
    <xsl:value-of select="@pin"/>
    <xsl:text>&nl;</xsl:text>
</xsl:template>
```

**</xsl:stylesheet>**

And here is the pads-pcb output file after running xsltproc:

```
*PADS-PCB*
*PART*
 P1 unknown
 U2 unknown
 U1 unknown
 C1 unknown
 R1 unknown

*NET*
*SIGNAL* GND
 U1.7
 C1.2
 U2.7
 P1.4
*SIGNAL* VCC
 R1.1
 U1.14
 U2.4
 U2.1
 U2.14
 P1.1
*SIGNAL* N-4
 U1.2
 U2.3
*SIGNAL* /SIG_OUT
 P1.2
 U2.5
 U2.2
*SIGNAL* /CLOCK_IN
 R1.2
 C1.1
 U1.1
 P1.3
*END*
```

The command line to make this conversion is:

kicad/bin/xsltproc.exe -o test.net kicad/bin/plugins/netlist_form_pads-pcb.xsl test.tmp

### 14.3.2 - Create a Cadstar netlist file

The Cadstar format is comprized of two sections.

- The footprint list.

- The Nets list: grouping pads references by nets.

Here is the style-sheet file to make this specific conversion:

```
<?xml version="1.0" encoding="ISO-8859-1"?>
<!--XSL style sheet to EESCHEMA Generic Netlist Format to CADSTAR netlist format
    Copyright (C) 2010, Jean-Pierre Charras.
    Copyright (C) 2010, SoftPLC Corporation.
    GPL v2.
-->

<!DOCTYPE xsl:stylesheet [
  <!ENTITY nl   "&#xd;&#xa;"> <!--new line CR, LF -->
]>

<xsl:stylesheet version="1.0" xmlns:xsl="http://www.w3.org/1999/XSL/Transform">
<xsl:output method="text" omit-xml-declaration="yes" indent="no"/>

<!-- Netlist header -->
<xsl:template match="/export">
    <xsl:text>.HEA&nl;</xsl:text>
    <xsl:apply-templates select="design/date"/>  <!-- Generate line .TIM <time> -->
    <xsl:apply-templates select="design/tool"/>  <!-- Generate line .APP <eeschema
version> -->
    <xsl:apply-templates select="components/comp"/>  <!-- Generate list of
components -->
    <xsl:text>&nl;&nl;</xsl:text>
    <xsl:apply-templates select="nets/net"/>          <!-- Generate list of nets and
connections -->
    <xsl:text>&nl;.END&nl;</xsl:text>
</xsl:template>

 <!-- Generate line .TIM 20/08/2010 10:45:33 -->
<xsl:template match="tool">
    <xsl:text>.APP "</xsl:text>
    <xsl:apply-templates/>
    <xsl:text>"&nl;</xsl:text>
</xsl:template>

 <!-- Generate line .APP "eeschema (2010-08-17 BZR 2450)-unstable" -->
<xsl:template match="date">
    <xsl:text>.TIM </xsl:text>
    <xsl:apply-templates/>
    <xsl:text>&nl;</xsl:text>
</xsl:template>

<!-- for each component -->
<xsl:template match="comp">
    <xsl:text>.ADD_COM </xsl:text>
    <xsl:value-of select="@ref"/>
    <xsl:text> </xsl:text>
    <xsl:choose>
        <xsl:when test = "value != '' ">
            <xsl:text>"</xsl:text> <xsl:apply-templates select="value"/>
<xsl:text>"</xsl:text>
        </xsl:when>
        <xsl:otherwise>
            <xsl:text>""</xsl:text>
        </xsl:otherwise>
    </xsl:choose>
```

```
    <xsl:text>&nl;</xsl:text>
</xsl:template>

<!-- for each net -->
<xsl:template match="net">
    <!-- nets are output only if there is more than one pin in net -->
    <xsl:if test="count(node)>1">
    <xsl:variable name="netname">
        <xsl:text>"</xsl:text>
        <xsl:choose>
            <xsl:when test = "@name != '' ">
                <xsl:value-of select="@name"/>
            </xsl:when>
            <xsl:otherwise>
                <xsl:text>N-</xsl:text>
                <xsl:value-of select="@code"/>
        </xsl:otherwise>
        </xsl:choose>
        <xsl:text>"&nl;</xsl:text>
        </xsl:variable>
        <xsl:apply-templates select="node" mode="first"/>
        <xsl:value-of select="$netname"/>
        <xsl:apply-templates select="node" mode="others"/>
    </xsl:if>
</xsl:template>

<!-- for each node -->
<xsl:template match="node" mode="first">
    <xsl:if test="position()=1">
        <xsl:text>.ADD_TER </xsl:text>
    <xsl:value-of select="@ref"/>
    <xsl:text>.</xsl:text>
    <xsl:value-of select="@pin"/>
    <xsl:text> </xsl:text>
    </xsl:if>
</xsl:template>

<xsl:template match="node" mode="others">
    <xsl:choose>
        <xsl:when test='position()=1'>
        </xsl:when>
        <xsl:when test='position()=2'>
            <xsl:text>.TER      </xsl:text>
        </xsl:when>
        <xsl:otherwise>
            <xsl:text>             </xsl:text>
        </xsl:otherwise>
    </xsl:choose>
    <xsl:if test="position()>1">
        <xsl:value-of select="@ref"/>
        <xsl:text>.</xsl:text>
        <xsl:value-of select="@pin"/>
        <xsl:text>&nl;</xsl:text>
    </xsl:if>
</xsl:template>

</xsl:stylesheet>
```

Here is the Cadstar output file.

```
.HEA
.TIM 21/08/2010 08:12:08
```

```
.APP "eeschema (2010-08-09 BZR 2439)-unstable"
.ADD_COM P1 "CONN_4"
.ADD_COM U2 "74LS74"
.ADD_COM U1 "74LS04"
.ADD_COM C1 "CP"
.ADD_COM R1 "R"


.ADD_TER U1.7 "GND"
.TER      C1.2
          U2.7
          P1.4
.ADD_TER R1.1 "VCC"
.TER      U1.14
          U2.4
          U2.1
          U2.14
          P1.1
.ADD_TER U1.2 "N-4"
.TER      U2.3
.ADD_TER P1.2 "/SIG_OUT"
.TER      U2.5
          U2.2
.ADD_TER R1.2 "/CLOCK_IN"
.TER      C1.1
          U1.1
          P1.3


.END
```

### 14.3.3 - Create a OrcadPCB2 netlist file

This format has only one section which is the footprint list. Each footprint includes its list of pads with reference to a net.

Here is the style-sheet for this specific conversion:

```
<?xml version="1.0" encoding="ISO-8859-1"?>
<!--XSL style sheet to EESCHEMA Generic Netlist Format to CADSTAR netlist format
    Copyright (C) 2010, SoftPLC Corporation.
    GPL v2.

    How to use:
       https://lists.launchpad.net/kicad-developers/msg05157.html
-->

<!DOCTYPE xsl:stylesheet [
  <!ENTITY nl  "&#xd;&#xa;"> <!--new line CR, LF -->
]>

<xsl:stylesheet version="1.0" xmlns:xsl="http://www.w3.org/1999/XSL/Transform">
<xsl:output method="text" omit-xml-declaration="yes" indent="no"/>

<!--
    Netlist header
    Creates the entire netlist
    (can be seen as equivalent to main function in C
-->
<xsl:template match="/export">
    <xsl:text>( { EESchema Netlist Version 1.1  </xsl:text>
    <!-- Generate line .TIM <time> -->
    <xsl:apply-templates select="design/date"/>
    <!-- Generate line eeschema version ... -->
    <xsl:apply-templates select="design/tool"/>
```

```xml
<xsl:text>}&nl;</xsl:text>

<!-- Generate the list of components -->
<xsl:apply-templates select="components/comp"/>  <!-- Generate list of components -->

<!-- end of file -->
<xsl:text>)&nl;*&nl;</xsl:text>
</xsl:template>

<!--
  Generate id in header like "eeschema (2010-08-17 BZR 2450)-unstable"
-->
<xsl:template match="tool">
  <xsl:apply-templates/>
</xsl:template>

<!--
  Generate date in header like "20/08/2010 10:45:33"
-->
<xsl:template match="date">
  <xsl:apply-templates/>
  <xsl:text>&nl;</xsl:text>
</xsl:template>

<!--
  This template read each component
  (path = /export/components/comp)
  creates lines:
   ( 3EBF7DBD $noname U1 74LS125
    ... pin list ...
    )
  and calls "create_pin_list" template to build the pin list
-->
<xsl:template match="comp">
  <xsl:text> ( </xsl:text>
  <xsl:choose>
    <xsl:when test = "tstamp != '' ">
      <xsl:apply-templates select="tstamp"/>
    </xsl:when>
    <xsl:otherwise>
      <xsl:text>00000000</xsl:text>
    </xsl:otherwise>
  </xsl:choose>
  <xsl:text> </xsl:text>
  <xsl:choose>
    <xsl:when test = "footprint != '' ">
      <xsl:apply-templates select="footprint"/>
    </xsl:when>
    <xsl:otherwise>
      <xsl:text>$noname</xsl:text>
    </xsl:otherwise>
  </xsl:choose>
  <xsl:text> </xsl:text>
  <xsl:value-of select="@ref"/>
  <xsl:text> </xsl:text>
  <xsl:choose>
    <xsl:when test = "value != '' ">
      <xsl:apply-templates select="value"/>
    </xsl:when>
    <xsl:otherwise>
      <xsl:text>"~"</xsl:text>
```

```
          </xsl:otherwise>
        </xsl:choose>
        <xsl:text>&nl;</xsl:text>
        <xsl:call-template name="Search_pin_list" >
          <xsl:with-param name="cmplib_id" select="libsource/@part"/>
          <xsl:with-param name="cmp_ref" select="@ref"/>
        </xsl:call-template>
        <xsl:text> )&nl;</xsl:text>
</xsl:template>


<!--
    This template search for a given lib component description in list
    lib component descriptions are in /export/libparts,
    and each description start at ./libpart
    We search here for the list of pins of the given component
    This template has 2 parameters:
        "cmplib_id" (reference in libparts)
        "cmp_ref"   (schematic reference of the given component)
-->
<xsl:template name="Search_pin_list" >
    <xsl:param name="cmplib_id" select="0" />
    <xsl:param name="cmp_ref" select="0" />
        <xsl:for-each select="/export/libparts/libpart">
            <xsl:if test = "@part = $cmplib_id ">
                <xsl:apply-templates name="build_pin_list" select="pins/pin">
                    <xsl:with-param name="cmp_ref" select="$cmp_ref"/>
                </xsl:apply-templates>
            </xsl:if>
        </xsl:for-each>
</xsl:template>



<!--
    This template writes the pin list of a component
    from the pin list of the library description
    The pin list from library description is something like
        <pins>
        <pin num="1" type="passive"/>
        <pin num="2" type="passive"/>
        </pins>
    Output pin list is ( <pin num> <net name> )
    something like
        ( 1 VCC )
        ( 2 GND )
-->
<xsl:template name="build_pin_list" match="pin">
    <xsl:param name="cmp_ref" select="0" />

    <!-- write pin numner and separator -->
    <xsl:text> ( </xsl:text>
    <xsl:value-of select="@num"/>
    <xsl:text> </xsl:text>

    <!-- search net name in nets section and write it: -->
    <xsl:variable name="pinNum" select="@num" />
    <xsl:for-each select="/export/nets/net">
        <!-- net name is output only if there is more than one pin in net
             else use "?" as net name, so count items in this net
        -->
        <xsl:variable name="pinCnt" select="count(node)" />
        <xsl:apply-templates name="Search_pin_netname" select="node">
```

```
          <xsl:with-param name="cmp_ref" select="$cmp_ref"/>
          <xsl:with-param name="pin_cnt_in_net" select="$pinCnt"/>
          <xsl:with-param name="pin_num"> <xsl:value-of select="$pinNum"/>
          </xsl:with-param>
      </xsl:apply-templates>
   </xsl:for-each>


   <!-- close line -->
   <xsl:text> )&nl;</xsl:text>
</xsl:template>

<!--
   This template writes the pin netname of a given pin of a given component
   from the nets list
   The nets list description is something like
     <nets>
       <net code="1" name="GND">
        <node ref="J1" pin="20"/>
          <node ref="C2" pin="2"/>
      </net>
       <net code="2" name="">
        <node ref="U2" pin="11"/>
       </net>
    </nets>
   This template has 2 parameters:
      "cmp_ref"  (schematic reference of the given component)
      "pin_num"   (pin number)
-->

<xsl:template name="Search_pin_netname" match="node">
   <xsl:param name="cmp_ref" select="0" />
   <xsl:param name="pin_num" select="0" />
   <xsl:param name="pin_cnt_in_net" select="0" />

   <xsl:if test = "@ref = $cmp_ref ">
      <xsl:if test = "@pin = $pin_num">
      <!-- net name is output only if there is more than one pin in net
         else use "?" as net name
      -->
         <xsl:if test = "$pin_cnt_in_net>1">
           <xsl:choose>
              <!-- if a net has a name, use it,
                 else build a name from its net code
              -->
              <xsl:when test = "../@name != '' ">
                 <xsl:value-of select="../@name"/>
              </xsl:when>
              <xsl:otherwise>
                 <xsl:text>$N-0</xsl:text><xsl:value-of select="../@code"/>
              </xsl:otherwise>
           </xsl:choose>
         </xsl:if>
         <xsl:if test = "$pin_cnt_in_net &lt;2">
            <xsl:text>?</xsl:text>
         </xsl:if>
      </xsl:if>
   </xsl:if>

</xsl:template>

</xsl:stylesheet>
```

Here is the OrcadPCB2 output file.

```
( { EESchema Netlist Version 1.1  29/08/2010 21:07:51
eeschema (2010-08-28 BZR 2458)-unstable}
 ( 4C6E2141 $noname P1 CONN_4
  ( 1 VCC )
  ( 2 /SIG_OUT )
  ( 3 /CLOCK_IN )
  ( 4 GND )
 )
 ( 4C6E20BA $noname U2 74LS74
  ( 1 VCC )
  ( 2 /SIG_OUT )
  ( 3 N-04 )
  ( 4 VCC )
  ( 5 /SIG_OUT )
  ( 6 ? )
  ( 7 GND )
  ( 14 VCC )
 )
 ( 4C6E20A6 $noname U1 74LS04
  ( 1 /CLOCK_IN )
  ( 2 N-04 )
  ( 7 GND )
  ( 14 VCC )
 )
 ( 4C6E2094 $noname C1 CP
  ( 1 /CLOCK_IN )
  ( 2 GND )
 )
 ( 4C6E208A $noname R1 R
  ( 1 VCC )
  ( 2 /CLOCK_IN )
 )
)
*
```

### 14.3.4 - Eeschema plugins interface

Intermediate Netlist converters can be automatically launched within Eeschema.

### 14.3.4.1 - Init the Dialog window

One can add a new netlist plug-in user interface tab by clicking on the Add Plugin tab.

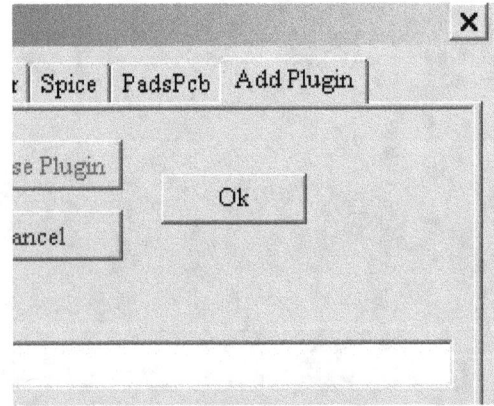

Here is what the configuration data for the PadsPcb tab looks like:

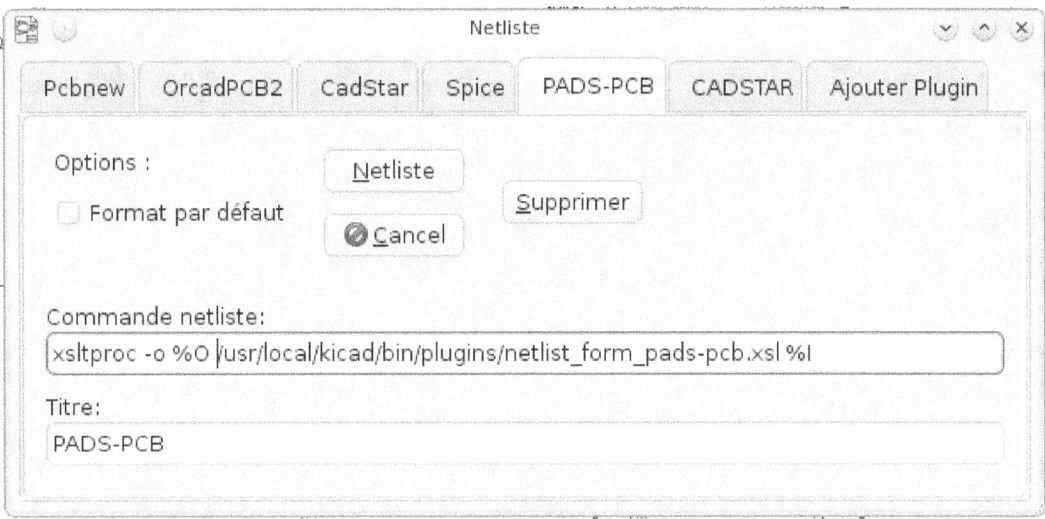

### 14.3.4.2 - Plugin Configuration Parameters

The Eeschema plug-in configuration dialog requires the following information:

- The title: for instance, the name of the netlist format.
- The command line to launch the converter.

Once you click on the netlist button the following will happen:

1. Eeschema creates an intermediate netlist file *.xml, for instance *test.xml.*
2. Eeschema runs the plug-in by reading test.xml and creates test.net

### 14.3.4.3 - Generate netlist files with the command line

Assuming we are using the program *xsltproc.exe* to apply the sheet style to the intermediate file, *xsltproc.exe* is executed with the following command.

*xsltproc.exe -o <output filename> < style-sheet filename> <input XML file to convert>*

In Kicad under Windows the command line is the following.

*f:/kicad/bin/xsltproc.exe -o "%O" f:/kicad/bin/plugins/netlist_form_pads-pcb.xsl "%I"*

Under Linux the command becomes as following.

*xsltproc -o "%O" /usr/local/kicad/bin/plugins/netlist_form_pads-pcb.xsl "%I"*

Where *netlist_form_pads-pcb.xsl* is the style-sheet that you are applying. Do not forget the double quotes around the file names, this allows them to have spaces after the substitution by Eeschema.

The command line format accepts parameters for filenames:

The supported formatting parameters are.

- %B => base filename and path of selected output file, minus path and extension.
- %I => complete filename and path of the temporary input file (the intermediate net file).
- %O => complete filename and path of the user chosen output file.

%I will be replaced by the actual intermediate file name

%O will be replaced by the actual output file name.

### 14.3.4.4 - Command line format: example for xsltproc

The command line format for  xsltproc is the following:

\<path of xsltproc> xsltproc \<xsltproc parameters>

*under Windows.*

**f:/kicad/bin/xsltproc.exe -o "%O" f:/kicad/bin/plugins/netlist_form_pads-pcb.xsl "%I"**

under Linux:

**xsltproc -o "%O" /usr/local/kicad/bin/plugins/netlist_form_pads-pcb.xsl "%I"**

The above examples assume xsltproc is installed on your PC under Windows and all files located in kicad/bin.

### *14.3.5 - Bill of Materials Generation*

Because the intermediate netlist file contains all information about used components, a BOM can be extracted from it. Here is the plug-in setup window (on Linux) to create a customized Bill Of Materials (BOM) file:

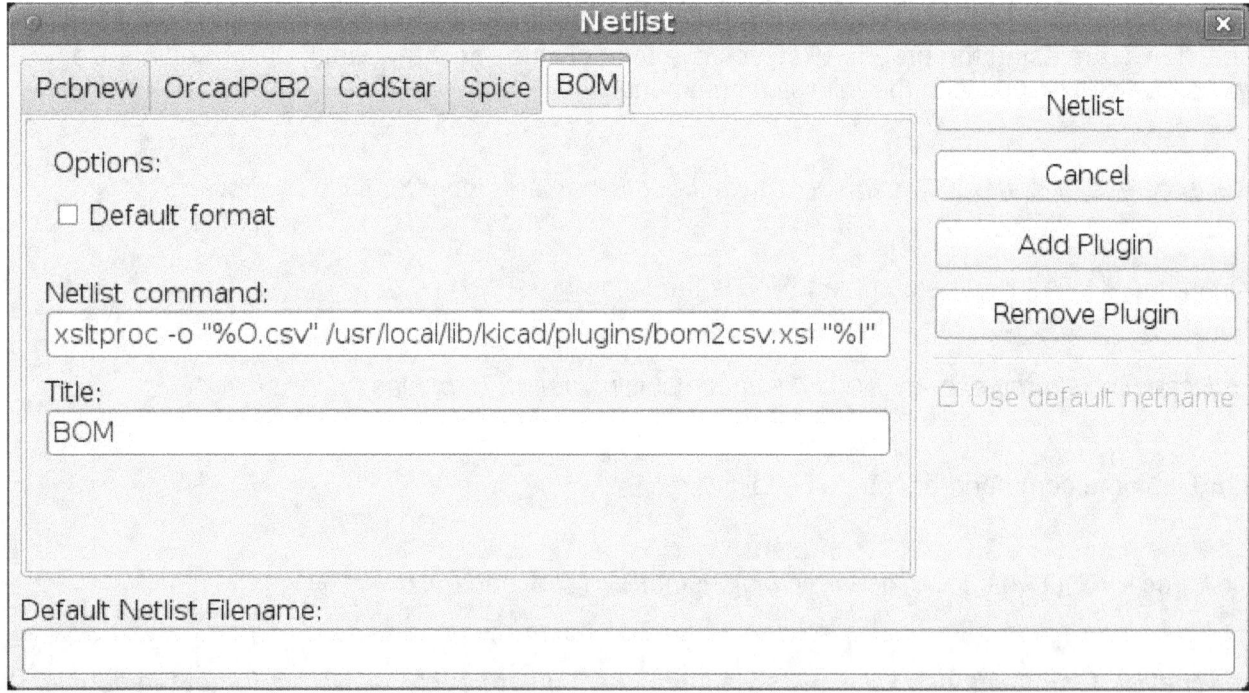

The path to the style sheet bom2csv.xsl is system dependent.  The currently best XSLT style-sheet for BOM generation at this time is called **bom2csv.xsl**. You are free to modify it according to your needs, and if you develop something generally useful, ask that it become part of the KiCad project.

## 14.4 - Command line format: example for python scripts

 The command line format for  python is something like:

python \<script file name> \<input filename> \<output filename>

*under Windows:*

**python.exe f:/kicad/python/my_python_script.py "%I" "%O"**

under Linux:

**python /usr/local/kicad/python/my_python_script.py "%I" "%O"**

Assuming python is installed on your PC.

## 14.5 - Intermediate Netlist structure

This sample gives an idea of the netlist file format.

```xml
<?xml version="1.0" encoding="utf-8"?>
<export version="D">
  <design>
    <source>F:\kicad_aux\netlist_test\netlist_test.sch</source>
    <date>29/08/2010 21:07:51</date>
    <tool>eeschema (2010-08-28 BZR 2458)-unstable</tool>
  </design>
  <components>
    <comp ref="P1">
      <value>CONN_4</value>
      <libsource lib="conn" part="CONN_4"/>
      <sheetpath names="/" tstamps="/"/>
      <tstamp>4C6E2141</tstamp>
    </comp>
    <comp ref="U2">
      <value>74LS74</value>
      <libsource lib="74xx" part="74LS74"/>
      <sheetpath names="/" tstamps="/"/>
      <tstamp>4C6E20BA</tstamp>
    </comp>
    <comp ref="U1">
      <value>74LS04</value>
      <libsource lib="74xx" part="74LS04"/>
      <sheetpath names="/" tstamps="/"/>
      <tstamp>4C6E20A6</tstamp>
    </comp>
    <comp ref="C1">
      <value>CP</value>
      <libsource lib="device" part="CP"/>
      <sheetpath names="/" tstamps="/"/>
      <tstamp>4C6E2094</tstamp>
    <comp ref="R1">
      <value>R</value>
      <libsource lib="device" part="R"/>
      <sheetpath names="/" tstamps="/"/>
      <tstamp>4C6E208A</tstamp>
    </comp>
  </components>
  <libparts/>
  <libraries/>
  <nets>
    <net code="1" name="GND">
      <node ref="U1" pin="7"/>
      <node ref="C1" pin="2"/>
      <node ref="U2" pin="7"/>
      <node ref="P1" pin="4"/>
    </net>
    <net code="2" name="VCC">
      <node ref="R1" pin="1"/>
      <node ref="U1" pin="14"/>
      <node ref="U2" pin="4"/>
      <node ref="U2" pin="1"/>
      <node ref="U2" pin="14"/>
      <node ref="P1" pin="1"/>
    </net>
```

```
          <net code="3" name="">
            <node ref="U2" pin="6"/>
          </net>
          <net code="4" name="">
            <node ref="U1" pin="2"/>
            <node ref="U2" pin="3"/>
          </net>
          <net code="5" name="/SIG_OUT">
            <node ref="P1" pin="2"/>
            <node ref="U2" pin="5"/>
            <node ref="U2" pin="2"/>
          </net>
          <net code="6" name="/CLOCK_IN">
            <node ref="R1" pin="2"/>
            <node ref="C1" pin="1"/>
            <node ref="U1" pin="1"/>
            <node ref="P1" pin="3"/>
          </net>
        </nets>
</export>
```

### 14.5.1 - General netlist file structure

The intermediate Netlist accounts for five sections.

- The header section.

- The component section.

- The lib parts section.

- The libraries section.

- The nets section.

The file content has the delimiter <export>

```
<export version="D">
  ...
</export>
```

### 14.5.2 - The header section

The header has the delimiter <design>

```
  <design>
    <source>F:\kicad_aux\netlist_test\netlist_test.sch</source>
    <date>21/08/2010 08:12:08</date>
    <tool>eeschema (2010-08-09 BZR 2439)-unstable</tool>
  </design>
```

This section can be considered a comment section.

### 14.5.3 - The components section

The component section has the delimiter <components>

```
  <components>
    <comp ref="P1">
      <value>CONN_4</value>
      <libsource lib="conn" part="CONN_4"/>
      <sheetpath names="/" tstamps="/"/>
      <tstamp>4C6E2141</tstamp>
    </comp>
  </components>
```

This section contains the list of components in your schematic. Each component is described like this:

```
<comp ref="P1">
  <value>CONN_4</value>
  <libsource lib="conn" part="CONN_4"/>
  <sheetpath names="/" tstamps="/"/>
  <tstamp>4C6E2141</tstamp>
</comp>
```

| libsource | name of the lib where this component was found. |
|---|---|
| **part** | component name inside this library. |
| **sheetpath** | path of the sheet inside the hierarchy: identify the sheet within the full schematic hierarchy. |
| **tstamps (time stamps)** | time stamp of the schematic file. |
| **tstamp (time stamp)** | time stamp of the component. |

### 14.5.3.1 - Note about time stamps for components

To identify a component in a netlist and therefore on a board, the timestamp reference is used as unique for each component.  However Kicad provides an auxiliary way to identify a component which is the corresponding footprint on the board. This allows the re-annotation of components in a schematic project and does not loose the link between the component and its footprint.

A time stamp is an unique identifier for each component or sheet in a schematic project. However, in complex hierarchies, the same sheet is used more than once, so this sheet contains components having the same time stamp.

A given sheet inside a complex hierarchy has an unique identifier: its sheetpath. A given component (inside a complex hierarchy) has an unique identifier: the sheetpath + its tstamp

### *14.5.4 - The libparts section*

The libparts section has the delimiter <libparts>, and the content of this section is defined in the schematic libraries. The libparts section contains

- The allowed footprints names (names use jokers) delimiter <fp>.

- The fields defined in the library delimiter <fields>.

- The list of pins delimiter <pins>.

```
<libparts>
  <libpart lib="device" part="CP">
    <description>Condensateur polarise</description>
    <footprints>
      <fp>CP*</fp>
      <fp>SM*</fp>
    </footprints>
    <fields>
      <field name="Reference">C</field>
      <field name="Valeur">CP</field>
    </fields>
    <pins>
      <pin num="1" name="1" type="passive"/>
      <pin num="2" name="2" type="passive"/>
    </pins>
  </libpart>
</libparts>
```

Lines like `<pin num="1" type="passive"/>` give also the electrical pin type. Possible electrical pin types are

| Input | Usual input pin |
|---|---|
| Output | Usual output |
| Bidirectional | Input or Output |
| Tri-state | Bus input/output |
| Passive | Usual ends of passive components |
| Unspecified | Unknown electrical type |
| Power input | Power input of a component |
| Power output | Power output like a regulator output |
| Open collector | Open collector often found in analog comparators |
| Open emitter | Open collector sometimes found in logic. |
| Not connected | Must be left open in schematic |

## 14.5.5 - The libraries section

The libraries section has the delimiter <libraries>. This section contains the list of schematic libraries used in the project.

```
<libraries>
  <library logical="device">
    <uri>F:\kicad\share\library\device.lib</uri>
  </library>
  <library logical="conn">
    <uri>F:\kicad\share\library\conn.lib</uri>
  </library>
</libraries>
```

## 14.5.6 - The nets section

The nets section has the delimiter <nets>. This section contains the "connectivity" of the schematic.

```
<nets>
  <net code="1" name="GND">
    <node ref="U1" pin="7"/>
    <node ref="C1" pin="2"/>
    <node ref="U2" pin="7"/>
    <node ref="P1" pin="4"/>
  </net>
  <net code="2" name="VCC">
    <node ref="R1" pin="1"/>
    <node ref="U1" pin="14"/>
    <node ref="U2" pin="4"/>
    <node ref="U2" pin="1"/>
    <node ref="U2" pin="14"/>
    <node ref="P1" pin="1"/>
  </net>
</nets>
```

This section lists all nets in the schematic.

A possible net is contains the following.

```
<net code="1" name="GND">
  <node ref="U1" pin="7"/>
  <node ref="C1" pin="2"/>
  <node ref="U2" pin="7"/>
  <node ref="P1" pin="4"/>
</net>
```

| net code | is an internal identifier for this net |
|---|---|
| name | is a name for this net |
| node | give a pin reference connected to this net |

# 14.6 - More about xsltproc

Refer to the page: *http://xmlsoft.org/XSLT/xsltproc.html*

## 14.6.1 - Introduction

xsltproc is a command line tool for applying XSLT style-sheets to XML documents. While it was developed as part of the GNOME project, it can operate independently of the GNOME desktop.

xsltproc is invoked from the command line with the name of the style-sheet to be used followed by the name of the file or files to which the style-sheet is to be applied. It will use the standard input if a filename provided is - .

If a style-sheet is included in an XML document with a Style-sheet Processing Instruction, no style-sheet needs to be named in the command line. xsltproc will automatically detect the included style-sheet and use it. By default, the output is to *stdout*. You can specify a file for output using the -o option.

## 14.6.2 - Synopsis

xsltproc [[-V] | [-v] | [-o *file*] | [--timing] | [--repeat] | [--debug] | [--novalid] | [--noout] | [--maxdepth *val*] | [--html] | [--param *name value*] | [--stringparam *name value*] | [--nonet] | [--path *paths*] | [--load-trace] | [--catalogs] | [--xinclude] | [--profile] | [--dumpextensions] | [--nowrite] | [--nomkdir] | [--writesubtree] | [--nodtdattr]] [*stylesheet*] [*file1*] [*file2*] [....]

## 14.6.3 - Command line options

*-V or --version*

Show the version of libxml and libxslt used.

*-v or --verbose*

Output each step taken by xsltproc in processing the stylesheet and the document.

*-o or --output file*

Direct output to the file named *file*. For multiple outputs, also known as "chunking", -o directory/ directs the output files to a specified directory. The directory must already exist.

*--timing*

Display the time used for parsing the stylesheet, parsing the document and applying the stylesheet and saving the result. Displayed in milliseconds.

*--repeat*

Run the transformation 20 times. Used for timing tests.

*--debug*

Output an XML tree of the transformed document for debugging purposes.

*--novalid*

Skip loading the document's DTD.

*--noout*

Do not output the result.

*--maxdepth value*

Adjust the maximum depth of the template stack before libxslt concludes it is in an infinite loop. The default is 500.

*--html*

The input document is an HTML file.

*--param name value*

Pass a parameter of name *name* and value *value* to the stylesheet. You may pass multiple name/value pairs up to a maximum of 32. If the value being passed is a string rather than a node identifier, use --stringparam instead.

*--stringparam name value*

Pass a paramenter of name *name* and value *value* where *value* is a string rather than a node identifier. (Note: The string must be utf-8.)

*--nonet*

Do not use the Internet to fetch DTD's, entities or documents.

*--path paths*

Use the list (separated by space or column) of filesystem paths specified by *paths* to load DTDs, entities or documents.

*--load-trace*

Display to stderr all the documents loaded during the processing.

*--catalogs*

Use the SGML catalog specified in SGML_CATALOG_FILES to resolve the location of external entities. By default, xsltproc looks for the catalog specified in XML_CATALOG_FILES. If that is not specified, it uses /etc/xml/catalog.

*--xinclude*

Process the input document using the Xinclude specification. More details on this can be found in the Xinclude specification: *http://www.w3.org/TR/xinclude/*

*--profile or --norman*

Output profiling information detailing the amount of time spent in each part of the stylesheet. This is useful in optimizing stylesheet performance.

*--dumpextensions*

Dumps the list of all registered extensions to stdout.

*--nowrite*

Refuses to write to any file or resource.

*--nomkdir*

Refuses to create directories.

*--writesubtree path*

Allow file write only within the *path* subtree.

*--nodtdattr*

Do not apply default attributes from the document's DTD.

### 14.6.4 - Xsltproc return values

xsltproc returns a status number that can be quite useful when calling it within a script.

0: normal

1: no argument

2: too many parameters

3: unknown option

4: failed to parse the stylesheet

5: error in the stylesheet

6: error in one of the documents

7: unsupported xsl:output method

8: string parameter contains both quote and double-quotes

9: internal processing error

10: processing was stopped by a terminating message

11: could not write the result to the output file

### 14.6.5 - More Information about xsltproc

libxml web page: *http://www.xmlsoft.org/*

W3C XSLT page: *http://www.w3.org/TR/xslt*

www.ingramcontent.com/pod-product-compliance
Lightning Source LLC
Chambersburg PA
CBHW082110220526
45472CB00009B/2116